LIFE SCIENCES CONTRIBUTIONS
ROYAL ONTARIO MUSEUM
NUMBER 103

JOHN E. STORER

Tertiary Mammals of Saskatchewan Part III: The Miocene Fauna

Publication date: 21 March 1975

ISBN 0-88854-168-6

Suggested citation: Life Sci. Contr., R. Ont. Mus.

ROYAL ONTARIO MUSEUM

PUBLICATIONS IN LIFE SCIENCES

The Royal Ontario Museum publishes three series in the Life Sciences:

LIFE SCIENCES CONTRIBUTIONS, a numbered series of original scientific publications, including monographic works.

LIFE SCIENCES OCCASIONAL PAPERS, a numbered series of original scientific publications, primarily short and usually of taxonomic significance.

LIFE SCIENCES MISCELLANEOUS PUBLICATIONS, an unnumbered series of publications of varied subject matter and format.

All manuscripts considered for publication are subject to the scrutiny and editorial policies of the Life Sciences Editorial Board, and to review by persons outside the Museum staff who are authorities in the particular field involved.

JOHN E. STORER is Curator of Paleontology at the Provincial Museum of Alberta, Edmonton. This publication is based on research carried out in the Department of Vertebrate Palaeontology and Office of the Chief Biologist, Royal Ontario Museum, for which he was awarded a doctorate in the University of Toronto, Department of Zoology.

PRICE: $4.00

100 Queen's Park, Toronto, Canada

PRINTED AND BOUND IN CANADA AT THE UNIVERSITY OF TORONTO PRESS

Contents

Tertiary Mammals of Saskatchewan Part III: The Miocene Fauna

Abstract

Fluvial sands and gravels of the Upper Miocene Wood Mountain Formation cap many high plateaux in southern Saskatchewan. The formation is considerably older than the nearby Flaxville complex of Montana and lies at higher elevation. All regions of the Wood Mountain Formation have yielded mammalian fossils of comparable ages, and no areas of major redeposition have been found.

The Wood Mountain mammalian fauna includes 57 species, including eight insectivores, twelve carnivores, one proboscidean, seven perissodactyls, ten artiodactyls, sixteen rodents, and three lagomorphs. Seven new species and two new genera of rodents and lagomorphs were named in 1970, and one new subspecies, *Copemys kelloggae canadensis*, is described here.

Analysis of the mammalian assemblage indicates that an age slightly greater than that of the Norden Bridge local fauna of Nebraska (basal Valentine Formation) can be estimated for the Wood Mountain Fauna. The fauna is thus uppermost Miocene.

(Mammalia; Miocene; Saskatchewan)

Introduction

Scope

This is the third contribution in a series on Tertiary mammals from Saskatchewan. Other reports include the Swift Current Creek Fauna, Upper Eocene (Russell, 1965), and the Cypress Hills Fauna, Lower Oligocene, non-ungulate orders (Russell, 1972).

Here is described the Upper Miocene mammalian fauna of the Wood Mountain Formation, the youngest Tertiary mammalian fauna known from Saskatchewan. I have studied fossil mammals from the Wood Mountain Formation in the National Museums of Canada, the Saskatchewan Museum of Natural History, and the Royal Ontario Museum. Collections made before 1969 from all localities are included here.

I have treated the geology of the formation briefly, as it is almost entirely the work of another author, Dr. J. A. Vonhof of the Groundwater Subdivision, Inland Waters Branch, Department of Energy, Mines and Resources, Government of Canada, Calgary (1965, 1969).

Geographical Extent and General Description of the Formation

Fluvial sands and gravels of the Wood Mountain Formation cap many high plateaux in southern Saskatchewan, in areas designated 72-G and 72-H in the Canadian National Topographic System, Canada Department of Mines and Technical Surveys, scales 1:50,000 and 1:250,000 (Wood Mountain Area and Willowbunch Lake Area). Exposure extends east from Twp 2, Rge 10, W. 3rd Merid. (49°08′ N, 107°14′ W), to Twp 1, Rge 23, W. 2nd Merid. (49°03′ N, 105°02′ W), and south from Twp 5, Rges 28 and 29, W. 2nd Merid. (49°23′ N, 105°38′–105°46′ W), well into Montana, perhaps as far south as Twp 34N, Rges 39 and 40E (48°39′ N, 106°33′ W; United States Department of the Interior, Geological Survey, National Topographic Map NM 13-10, Scale 1:250,000, Glasgow Sheet). I investigated only Saskatchewan occurrences.

The formation lies unconformably on the Ravenscrag Formation (buff Paleocene clays with coal seams) and rests on an erosional surface. Vonhof (1969, p. 126) estimated maximum thickness of the Wood Mountain Formation at 90–100 ft (27–31 m). Where uneroded, the gravels are usually at least 10 ft (3 m) thick.

To the west, the base of the formation lies at an altitude of 3,215 ft (980 m) in Twp 3, Rge 9, W. 3rd Merid. (49°12′ N, 107°10′ W; Vonhof, 1969, pp. 126–127), and to the east, southwest of Willowbunch, at approximately 2,875 ft (876 m). The altitude of the base of the Wood Mountain Formation increases from east to west, except in a large area of deposits (Twps 1 and 2, Rges 1, 2, and 3, W. 3rd Merid. 49°00′–49°10′ N, 106°02′–106°16′ W) which extends past the southern edge of the Wood Mountain Plateau and several miles south of the international boundary.

The nearby Flaxville complex of sands and gravels in Montana bears a superficial resemblance to the Wood Mountain Formation. It is considerably lower in elevation, and Collier and Thom (1918) stated that south of Redstone, Montana, it lies at an altitude of 2,600 ft (793 m). In its northern exposures, it is much younger than the Wood Mountain Formation (Storer, 1969).

Geology

Russell (1965) has given a general account of the Upper Cretaceous and Tertiary history of southern Saskatchewan. The Wood Mountain Formation bears many resemblances to the Cypress Hills Formation and, like that deposit, is the product of uplift in the Rocky Mountains and regional uplift in the plains, subsequent erosion, and deposition of a widespread sheet of cross-bedded sands and gravels in low-lying areas.

After the Upper Miocene, uplift took place once more in the Wood Mountain region, and stream erosion dissected the resultant plateau, reducing it to its present size. Glacial gravels are found on the northern edge of the plateau, but no glacial deposits occur on the top; thus most investigators have concluded that the northern edge of the plateau formed a barrier to the ice, and its top was never glaciated. All regional uplift and most subsequent erosion took place before the glaciation of southern Saskatchewan.

History of Exploration of Miocene Deposits

Many early geologists compared the Wood Mountain gravels with glacial gravels (Rose, 1916) or considered them equivalent in age to the Cypress Hills beds (Williams and Dyer, 1930). But in 1929, C. M. Sternberg (1930) discovered vertebrate fossils on the north slope of Wood Mountain and designated his locality, in Sect. 24, Twp 4, Rge 3, W. 3rd Merid. (49°19′ N, 106°18′ W), the type locality of the Wood Mountain Formation (Sternberg, op. cit.: *contra* Fraser et al., 1935, who stated that the type locality is in the Willowbunch area). The fossils were identified by G. G. Simpson, and a Middle or Upper Miocene age was assigned to the gravels (Sternberg, op. cit.).

In 1930 and 1931, F. J. McLearn and party obtained from the Wood Mountain Formation, among other specimens, the holotype and paratype of *Merychippus praecocidens* (Russell, 1933), herein referred to *Merychippus* cf. *M. isonesus.* Several subsequent expeditions searched for fossils in the Wood Mountain region (Russell, 1950). L. S. Russell (op. cit.) summarized all collections made prior to 1950.

In 1956, L. S. Russell and W. Langston, Jr. discovered the Yost Farm and Quantock Localities, two of the richest in the formation. Russell discovered the rich "Four Corners" locality in 1964. All three of these important localities are road cuts, and fossils from these and other road cuts throughout the area of exposure have been collected by workers from the National Museums of Canada, the Saskatchewan Museum of Natural History, and the Royal Ontario Museum. In 1967, L. S. Russell and I located a site on the farm of Mr. John Kleinfelder, near Rockglen, Saskatchewan, which has been highly productive of small mammalian material; Kleinfelder Farm specimens considered in this monograph were collected in 1967 and 1968.

Materials and Methods

Collections

Names of institutions housing material used for comparisons are abbreviated as follows:

AMNH	American Museum of Natural History, New York, New York (FAM Frick Collection)
CM	Carnegie Museum, Pittsburgh, Pennsylvania
GSC	Geological Survey of Canada (now part of NMC collections)
NMC	National Museum of Natural Sciences, National Museums of Canada, Ottawa, Ontario
ROM	Royal Ontario Museum, Vertebrate Palaeontology Collection, Toronto, Ontario
SMNH	Saskatchewan Museum of Natural History, Regina, Saskatchewan
UCMP	University of California, Museum of Paleontology, Berkeley, California
USNM	National Museum of Natural History, Washington, D.C.
YPM	Peabody Museum, Yale University, New Haven, Connecticut

SMNH numbers with prefix P are preliminary numbers; those without prefix are accession numbers. In all other instances, numbers referred to are catalogue numbers.

NMC and ROM catalogue numbers may include two or more specimens, so lists of specimens will show duplications. SMNH preliminary and accession numbers usually apply to collections containing two or more specimens, and a single number may appear in lists many times. When a single reference number applies to several specimens, they are boxed separately and labelled to avoid confusion.

The notation SMNH (?) applies to a single preliminary number at the Saskatchewan Museum of Natural History. Museum records include appropriate field data.

Specimens with prefix GSC are from collections of the Geological Survey of Canada, housed with collections of the National Museums of Canada. Roman numerals (e.g., GSC IV) are identifying numerals Sternberg put on his 1929 collection from the type locality. Specimens labelled GSC do not bear regular NMC catalogue numbers.

A mandible of *Monosaulax* from the private collection of Mr. Lawrence Yost of Rockglen, Saskatchewan, has been included in this study. A cast has been placed in the collections of the Royal Ontario Museum, but the specimen is noted as "L. Yost specimen" in the appropriate list, and measurements were taken on the original specimen.

A complete list of localities and the taxa collected from each appeared as Appendix B in Storer, 1970b, available on microfilm from the National

Library of Canada, Ottawa. Locality data are given only for type specimens in this paper.

Format

In descriptions of mammalian taxa, lower teeth, front to rear, are considered first, then lower jaw material, then upper teeth, front to rear, and then cranial material. This order is convenient, as Wood Mountain collections to date contain a preponderance of lower dental material over upper, and nearly all associated specimens are lower jaws rather than skulls. In palaeomerycids and antilocaprids, horns and horn cores are considered after identifiable teeth, as they belong with cranial material. Postcranial elements of taxonomic significance are described last, and most postcranial material in Wood Mountain collections is omitted.

Arrangement of orders and families of mammals follows Romer (1966, pp. 379–396). Genera are placed in an arbitrary sequence.

Terminology

A full diagram of dental terminology used in each group precedes its description. Groups treated are not taxonomically equivalent, so that groups represented by many species (e.g., selenodont Artiodactyla, Fig. 39) or one (e.g., Rhinocerotidae, Fig. 35) may share a dental terminology. In descriptions of mandibular, cranial, and postcranial material, I have used terminology in Romer (1966) or in standard texts on osteology.

Mensuration

Measurements of teeth of all insectivores, rodents, and lagomorphs except *Mylagaulus*, *Monosaulax*, *Amblycastor*, and *Hypolagus* were made with a Carl Zeiss measuring reticule (diameter 17 mm, scale 10 mm divided into 100 segments) in a Carl Zeiss 8X self-focusing eyepiece, and used on a Carl Zeiss "Student" microscope. An Ernst Leitz 3.5X objective was attached to the microscope giving an effective magnification of approximately 2.3X. When the microscope was focused on an object, a unit scale division represented .044 mm, and the entire scale represented 4.40 mm. Estimating one or two lines past the end of the scale, I could obtain a maximum accurate measurement of about 4.5 mm. Most objects measured on this microscope were 3 mm or smaller. Measurements taken on the microscope are expressed in units of 0.01 mm.

Teeth of the three rodents and the lagomorph mentioned above, plus teeth of carnivores, proboscideans, perissodactyls, and artiodactyls, were measured with vernier calipers, reading to 0.1 mm. All measurements are given in millimetres.

Except where noted, all measurements of teeth given in tables and lists are maxima, pertaining to the crown rather than to the occlusal surface, and anteroposterior and buccolingual (transverse) measurements of teeth are taken parallel to and perpendicular to the line of the tooth row respectively.

The following abbreviations are used in text, lists of measurements, and tables:

I incisor
C canine
dP deciduous (milk) premolar
P premolar
M molar
L left
R right
AP anteroposterior diameter
AP tri anteroposterior diameter of trigonid
BL buccolingual diameter
BL pro buccolingual diameter of protoloph or protolophid
BL met buccolingual diameter of metaloph or metalophid
BL hyp buccolingual diameter of hypoloph or hypolophid
BL tri buccolingual diameter of trigonid
BL tal buccolingual diameter of talonid
DV dorsoventral diameter
n number of specimens
OR observed range (smallest to largest)
CV coefficient of variation

Special measurements and measurements of the occlusal surface are explained where they occur. The mean, for samples of five or more, is given with standard error.

A list of measurements, arranged in paragraph form, is given for nearly every taxon. The object measured is given first, then the measurement, with number of specimens in parentheses, and mean, range, etc. For example, under *Hesperolagomys fluviatilis* (p. 111) appears P_3, AP (6), mean 1.41 ± .06, OR 1.19–1.63, CV 10.0; for third lower premolars, anteroposterior diameter was measured on six specimens, mean 1.41, standard error .06, observed range 1.19–1.63, coefficient of variation 10.0.

Illustrations

In photographs and line drawings, but not in diagrams illustrating terminology, a millimetre or centimetre scale appears near the bottom of each illustration, or, when all illustrations in a group are the same scale, at the bottom of the group.

In drawings except Figs. 20, 21, 29, and 30, cement is indicated by a pattern of x's, broken surfaces are stippled, and restored outlines of areas broken away are indicated by broken lines.

Systematic Description

Order Insectivora
Family Erinaceidae
Subfamily Brachyericinae

Brachyerix cf. *B. incertis* (Matthew), 1924

Fig. 2 A–C

REFERRED SPECIMENS. LM_2 (ROM 7633), RM_2 (ROM 7634, 7678), RM^1 (ROM 7632).

DESCRIPTION

M_2 (Fig. 2A, B) with strong entoconid-hypoconid, weak protoconid-hypoconid connections; hypoconid and entoconid closer together than protoconid and metaconid; no buccal cingulum at enamel margin of trigonid.

In M^1 (Fig. 2C), metacrista (ridge connecting hypocone and metaconule) directed more nearly posterad than in upper molars referred to *Parvericius* in this paper; posterobuccal corner of crown reduced, unlike that in *Parvericius*; large protocone joining paracone via high crest; low anterior cingulum running along anterior surface of crown, posterior cingulum along posterior enamel margin.

MEASUREMENTS

M_2, AP(2), OR 1.72–1.75; BL(2), OR 1.02–1.09. M^1, AP(1), 2.29; BL(1), 2.16. Measurements of M^1 not reliable, because anterobuccal corner of crown missing.

DISCUSSION

Second lower molars compare well in size with corresponding teeth of *Brachyerix incertis*, and topographic features are similar (Rich and Rich, 1971, table 3, fig. 18).

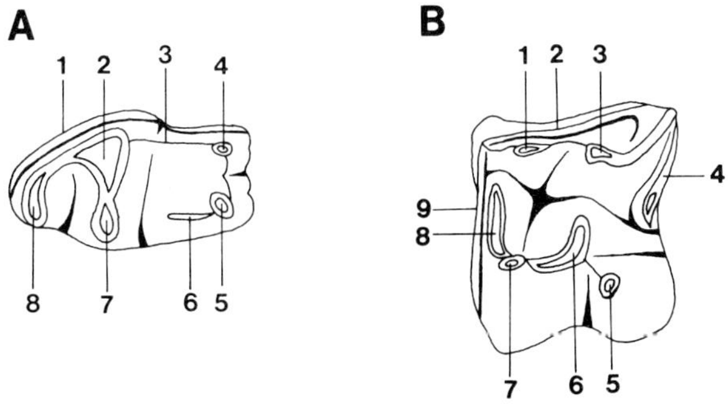

Fig. 1 Erinaceidae. Terminology used in description of molar teeth. Anterior to left, buccal to top of page. Not to scale.

A. RM_1 or M_2: 1, buccal cingulum; 2, protoconid; 3, hypolophid; 4, hypoconid; 5, entoconid; 6, entoconid ridge; 7, metaconid; 8, paraconid.

B. LM^1: 1, paracone; 2, buccal cingulum; 3, metacone; 4, posterior cingulum; 5, hypocone; 6, metaconule; 7, protocone; 8, protoconule; 9, anterior cingulum.

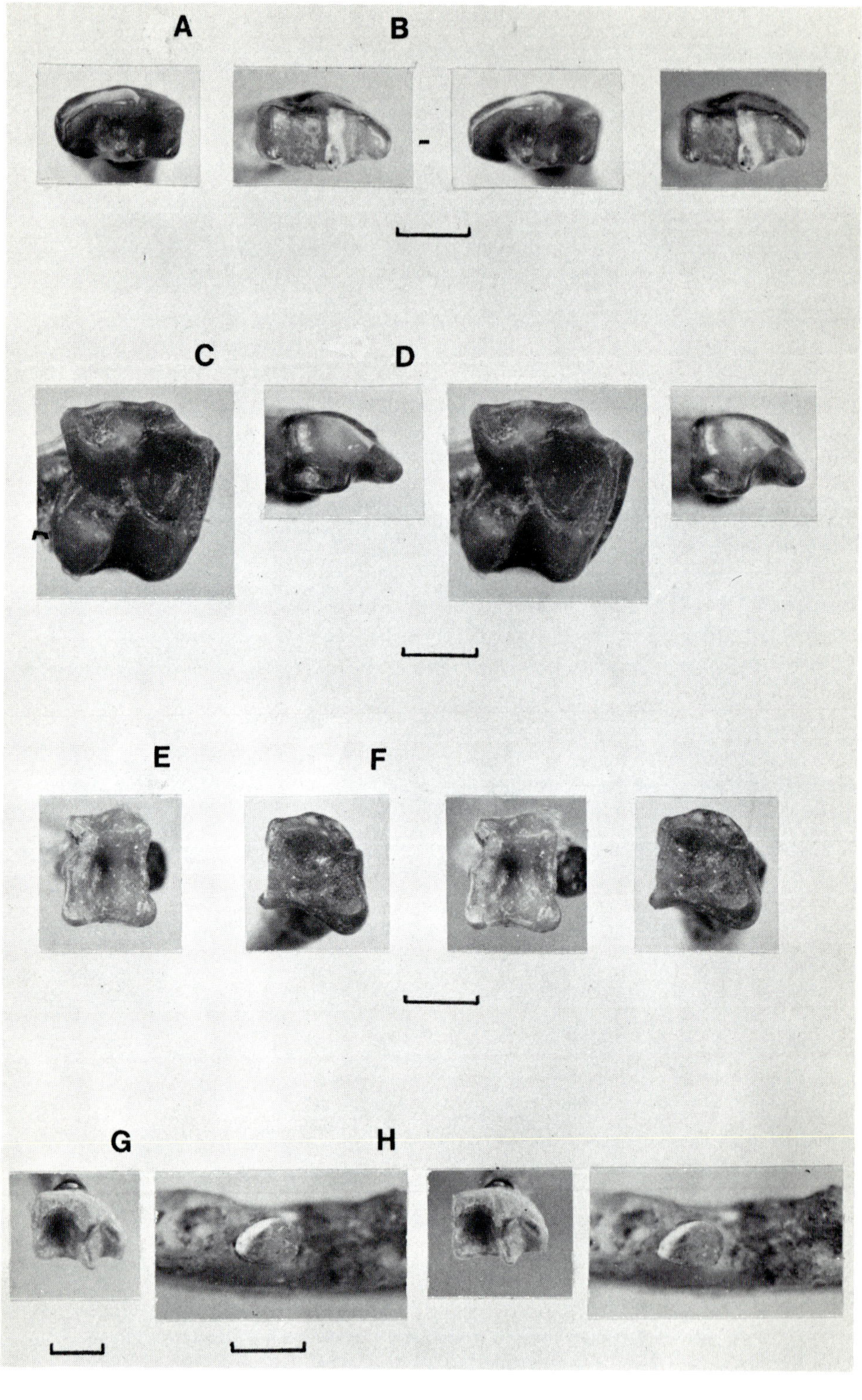
A
B
C
D
E
F
G
H

M^1, ROM 7632 differs, in the orientation of the metacrista, from M^1 of *Brachyerix macrotis* shown by Rich and Rich in fig. 9; in their specimen, the metacrista is directed more nearly posterad. Good agreement is shown with M^1 of *B. macrotis* in fig. 13, however. The degree of reduction of the posterobuccal corner compares well with *Brachyerix*. Although M^1 of *Brachyerix incertis* is not presently known, ROM 7632 is of the proper size to be referred to the same species as the lower molars. T. H. V. Rich (pers. comm., 1972) stated that this tooth shows a more nearly posteriorly directed metacrista than that in the type specimen of *Parvericius montanus* and is markedly smaller and more antero-posteriorly expanded relative to its width than M^1 of *Brachyerix macrotis*.

Subfamily Erinaceinae

Parvericius cf. *P. montanus* Koerner, 1940

Figs. 2 D–H; 3 A–C

REFERRED SPECIMENS. LP_4 (ROM 7671), RM_1 (ROM 7628, 7673), LM_2 (ROM 7624), R. mandibular fragment with RM_3 (ROM 7679), LM^1 (ROM 7629), LM^2 (ROM 7631), RM^2 (ROM 7630).

DESCRIPTION

P_4 (Fig. 2D) with large, distinct paraconid occupying anterior portion of crown; high ridge, apparently including small metaconid, running posterolinguad from protoconid to lingual border of crown; prominent ridge running from metaconid to posterolingual edge of crown; posterior shelf broad; anterobuccal cingulum occupying enamel margin beneath paraconid and protoconid; two roots.

Talonid of M_1 (Fig. 2E, F) with strong loph connecting entoconid and hypoconid, no strong hypolophid reaching forward from hypoconid.

M_2 (Fig. 2G) showing strong entoconid and entoconid crest; weak hypoconid weakly linked to entoconid; protoconid strongly connected to hypoconid by hypolophid; protoconid larger and higher than metaconid, and hypoconid and entoconid spaced about as widely as protoconid and metaconid.

M_3 (Fig. 2H) with no talonid; paraconid at anterolingual edge of crown; strong arcuate ridge extending to protoconid; protoconid and metaconid close together, strongly connected; short buccal cingulum anterobuccal to protoconid.

Ramus badly broken, cannot be described in detail. Mandibular depth beneath M_3, 2.3.

Fig. 2 *Brachyerix* cf. *B. incertis* and *Parvericius* cf. *P. montanus*, teeth in occlusal aspect. Scales 1 mm.

A-C. *Brachyerix* cf. *B. incertis*.

A. RM_2, ROM 7634, anterior to left, buccal to top of page.

B. LM_2, ROM 7633, anterior to right, buccal to top of page.

C. RM^1, ROM 7632, anterior to right, buccal to top of page.

D-H. *Parvericius* cf. *P. montanus*.

D. LP_4, ROM 7671, anterior to right, buccal to top of page.

E. RM_1, ROM 7628, anterior to left, buccal to top of page.

F. RM_1, ROM 7673, anterior to top of page, buccal to right.

G. LM_2, ROM 7624, anterior to right, buccal to top of page.

H. R. mandibular fragment with RM_3, ROM 7679, anterior to left, buccal to top of page.

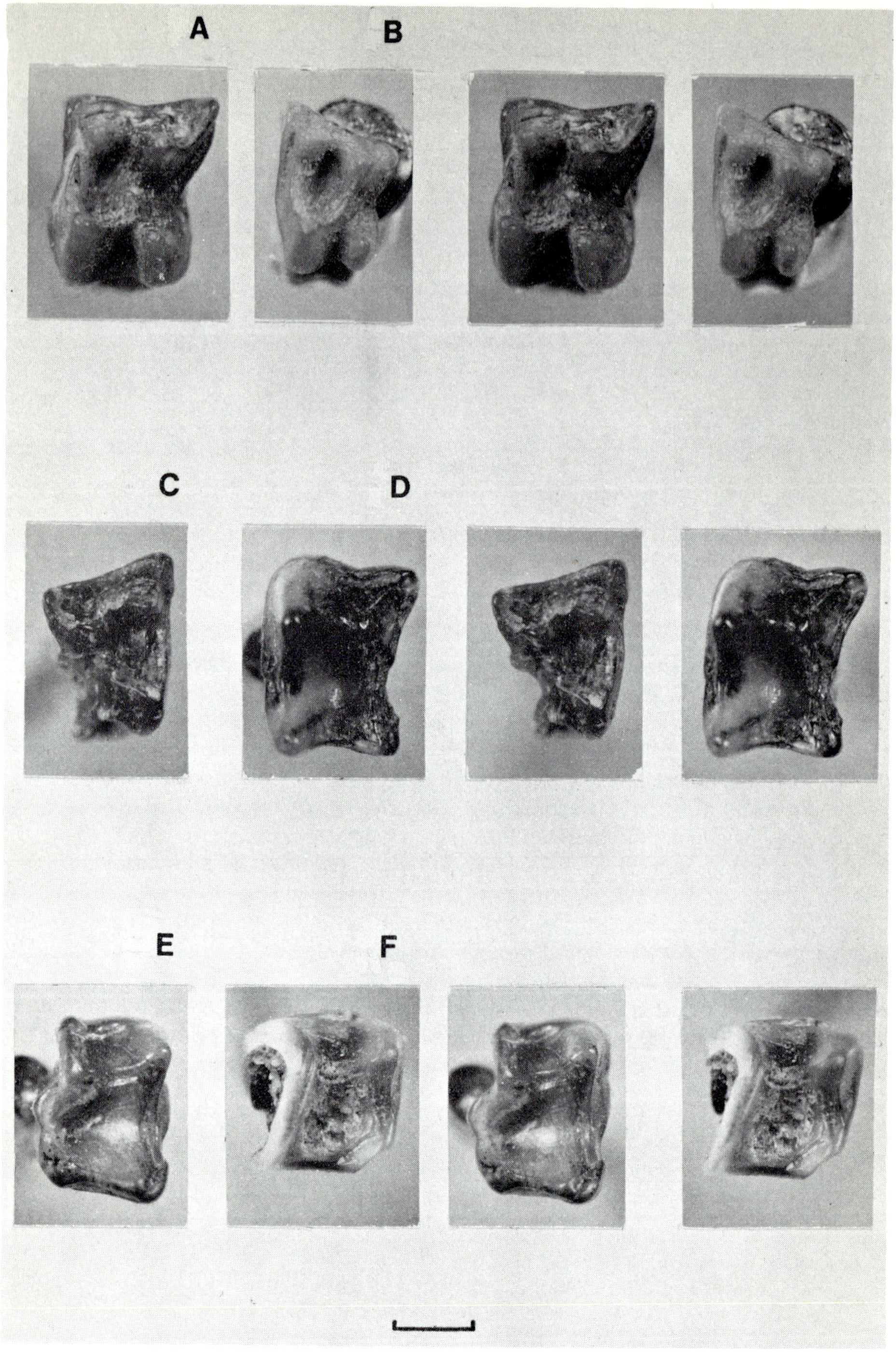

Fig. 3 *Parvericius* cf. *P. montanus* and *Amphechinus* sp., teeth in occlusal aspect. Buccal to top of page. Scale 1 mm.

A-C. *Parvericius* cf. *P. montanus.*

A. LM^1, ROM 7629, anterior to left.

B. LM^2, ROM 7631, anterior to left.

C. RM^2, ROM 7630, anterior to right.

D-F. *Amphechinus* sp.

D. LM_1, ROM 7627, anterior to right.

E. RM_1, ROM 7625, anterior to left.

F. RM_2, ROM 7672, anterior to left.

Upper molars (Fig. 3 A–C) showing large protocone joining paracone via high crest; lower anterior cingulum on anterior surface of crown; low metacrista joining metaconule and hypocone, running more nearly linguad than posterad.

Posterobuccal corner of M^1 unreduced; angle at anterobuccal corner of M^2 relatively great (T. H. V. Rich, pers. comm., 1972).

MEASUREMENTS

P_4, AP(1), 1.52; BL(1), 1.14. M_1, BLtal(2), OR 1.56–1.59. M_2, AP(1), 1.84; BL(1), 1.23. M_3, AP(1), 0.88; BL(1), 0.62. M^1, AP(1), 2.07; BL(1), 2.64. M^2, AP(2), OR 1.54–1.80; BL(1), 2.16.

DISCUSSION

T. H. V. Rich (pers. comm., 1972) has compared the Wood Mountain specimens with material at Yale University and the University of Michigan, and noted strong resemblance in size and general features of all but M^2 to *Parvericius montanus* Koerner. He stated concerning M^2 ROM 7630 and 7631, "If these teeth are those of a known North American erinaceid, they are the M^2 of *Parvericius* although differing from the type specimen of *P. montanus* in having a somewhat greater angle between the buccal and anterior sides."

Amphechinus sp.

Fig. 3 D–F

REFERRED SPECIMENS. LM_1 (ROM 7627), RM_1 (ROM 7625), RM_2 (ROM 7672).

DESCRIPTION

Talonids (Fig. 3 D–F) with strong loph connecting entoconid and hypoconid, no strong hypolophid reaching forward from hypoconid; larger than corresponding teeth referred to *Parvericius* cf. *P. montanus*.

MEASUREMENTS

M_1, BL tal (2), OR 2.23–2.42.

DISCUSSION

T. H. V. Rich (pers. comm., 1972) stated that the M_1 talonids agree in their large size with those of *Amphechinus horncloudi*, the largest known North American erinaceine. The talonid of M_2 is also very large and agrees in proportion of length to width with *Amphechinus*.

Family Talpidae

Domninoides sp.

FIG. 5A, B

REFERRED SPECIMENS. L. humerus (ROM 7635), R. humerus (ROM 7636).

DESCRIPTION

Humeri large, robust.

DISCUSSION

These humeri are larger and more robust than those referred to *Scalopoides* sp. (Fig. 7B, C). There are sufficient similarities to published illustrations of *Domninoides* (Reed, 1962; Hutchison, 1968) for reference to that genus.

No teeth which on the basis of size could reasonably be assigned to this taxon have been found.

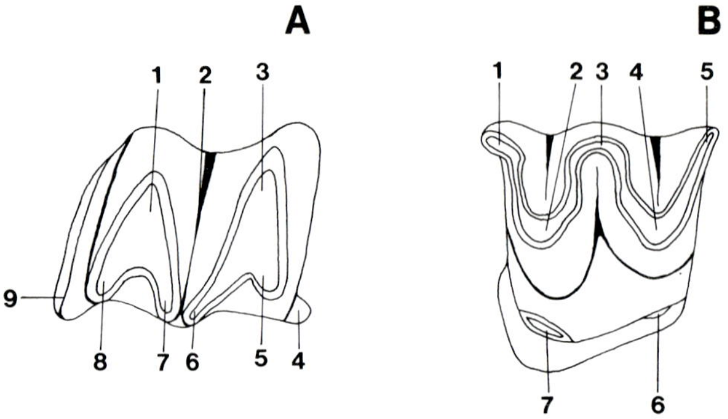

Fig. 4 Talpidae. Terminology used in description of molar teeth. Anterior to left, buccal to top of page. Not to scale.

A. RM_2: 1, protoconid; 2, hypoflexid; 3, hypoconid; 4, posterior accessory cuspule; 5, entoconid; 6, metastylid; 7, metaconid; 8, paraconid; 9, anterior cingulum.

B. LM^2: 1, parastyle; 2, paracone; 3, mesostyle; 4, metacone; 5, metastyle; 6, metaconule; 7, protocone.

Scalopoides sp.

FIGS. 5C, D; 6; 7

REFERRED SPECIMENS. LP_4 (ROM 7637), R. mandibular fragment with RP_4, RM_{1-2} (SMNH P384), LM_1 (ROM 7638, 7639), L. mandibular fragment with broken LM_1 (ROM 7689), RM_1 (ROM 7640, 7685, 7686), LM_2 (ROM 7641–7643, 7687), RM_2 (ROM 7644, 7645), R. mandibular fragment with broken RM_2 (ROM 7646), L. mandibular fragment with LM_{2-3} (ROM 7647), LM lower (ROM 7688), L. mandibular fragment (NMC 8988; ROM 7648, 7649), R. mandibular fragment (ROM 7650), LP^4 (ROM 7651, 7690), RP^4 (ROM 7652, 7653), LM^1 (ROM 7654, 7655), LM^2 (ROM 7656–7658), LM upper (ROM 7659), L. humerus (SMNH P551, P552; ROM 7660–7663), R. humerus (NMC 8976; specimen from Sternberg 1929 collection; ROM 7664, 7665).

DESCRIPTION

P_4 (Fig. 5C, D) with broader posterior shelf than in corresponding tooth of *Scalopoides isodens*; paraconid to the lingual side of line through accessory cuspule of posterior shelf, protoconid of P_4, and protoconids and hypoconids of M_1 and M_2, rather than on line, as described by Wilson (1960, p. 43) for *S. isodens*.

M_1 to M_3 (Figs. 5C, D; 6) much like corresponding teeth of *S. isodens*, but anterior arm of hypoconid of M_1 stronger, M_2 and M_3 with more prominent anterior cingula.

Lower jaw (SMNH P384, Fig. 5C, D) shaped as in other talpids; one mental foramen beneath P_3, another beneath septum between P_4 and M_1.

P^4 robust tritubercular tooth, with large, high paracone and distinct, laterally oriented metacone; protocone low but prominent, forming low basin; distinct anterior and posterior cingula; parastyle large.

M^1 long, angular; protocone and metaconule subequal, prominent; distinct anterior and posterior cingula present; parastyle large.

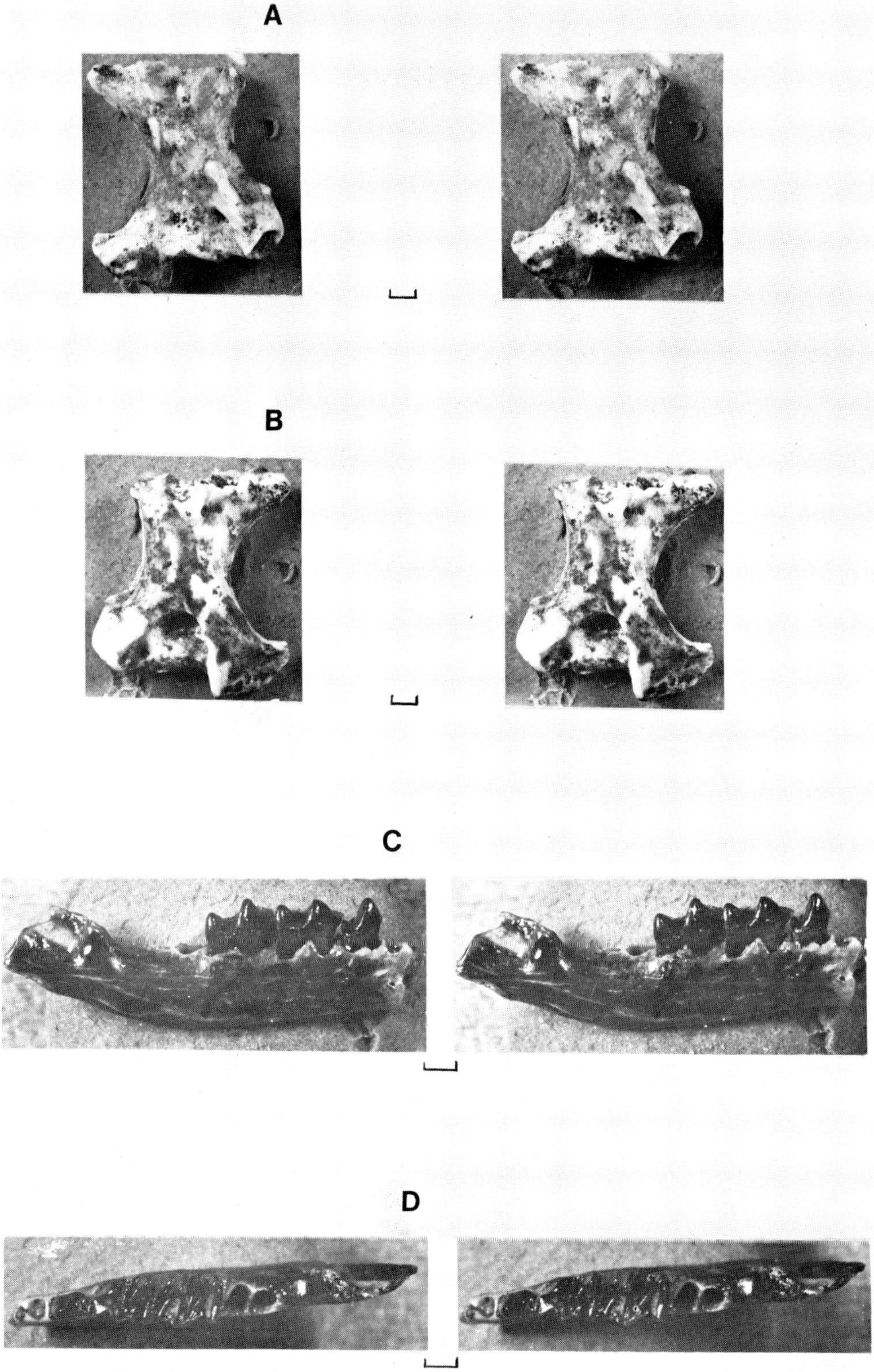

Fig. 5 *Domninoides* sp. and *Scalopoides* sp., humerus and mandibular ramus. Scales 1 mm.

A,B. *Domninoides* sp., R. humerus, ROM 7636, proximal to top of page.

A. Anterior aspect.

B. Posterior aspect.

C,D. *Scalopoides* sp., R. mandibular ramus with RP_4–RM_2, SMNH P384.

C. Lateral aspect, anterior to right, occlusal to top of page.

D. Occlusal aspect, anterior to left, buccal to top of page.

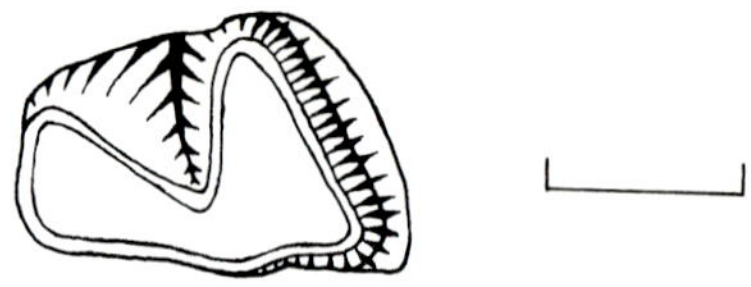

Fig. 6 *Scalopoides* sp., LM_3, ROM 7647, occlusal aspect. Anterior to right, buccal to top of page. Scale 1 mm.

M^2 (Fig. 7A) with small metaconule, large protocone; anterior and posterior cingula low and incomplete; parastyle large; tooth roughly rhomboidal in shape.

Humerus (Fig. 7B, C) similar in size and shape to those of *S. isodens* (Wilson, 1960, fig. 39), *Condylura cristata.*

MEASUREMENTS

P_4, AP(2), OR 1.06–1.54; BL(2), OR 0.77–0.88. M_1, AP(4), mean 2.16, OR 1.98–2.33; BL(4), mean 1.70, OR 1.61–1.85. M_2, AP(6), mean 2.50 ± .08, OR 2.22–2.76, CV 7.60; BL(6), mean 1.76 ± .06, OR 1.50–1.98, CV 8.49. M_3, AP(1), 2.02; BL(1), 1.50. P^4, AP(3), mean 1.89, OR 1.71–1.98; BL(3), mean 1.56, OR 1.45–1.63. *M^1, AP(1), 2.07; BL(1), 3.43. *M^2, AP(3), mean 2.19, OR 1.94–2.33; BL(3), mean 2.37, OR 2.18–2.64.

*Anteroposterior diameter of talpid upper molars taken perpendicular to anterior border of the crown, and buccolingual diameter taken parallel to anterior border.

DISCUSSION

The humeri suggest a talpid that was not as strongly fossorial as *Domninoides* sp.

This material is referable to *Scalopoides* on the basis of characteristics of lower cheek teeth and humeri. Further comparisons may indicate a new species of *Scalopoides.*

Talpidae, gen. et sp. indet.

FIG. 8

REFERRED SPECIMENS. RM_1 (ROM 7666), L. mandibular fragment (ROM 7681), LP^4 (ROM 7682), RP^4 (ROM 7683), RM^2 (ROM 7667, 7684), R. humerus (ROM 7668).

DESCRIPTION

M_1 (not shown) well worn; low anterior cingulum occupying enamel margin from paraconid to base of hypoconid; prominent posterior cingulum at posterior enamel margin; low stylid blocking ectoflexid.

P^4 (not shown) tritubercular; protocone and paracone joined by prominent loph running anterobuccad; protocone and metacone separated by persistent valley; shallow basin formed buccal to protocone; short cingulum occupying buccal enamel margin.

M^2 (Fig. 8A) with doubled protocone and prominent metaconule; parastyle small; central basin, buccal to protocone, deep.

Humerus (Fig. 8B, C) relatively long, slender.

MEASUREMENTS

M_1, AP(1), 1.17; BL(1), 1.08. P^4, AP(1), 1.43; BL(2), OR 1.10–1.21. *M^2, AP(2), OR 1.32–1.56; BL(1), 1.89.

*Measurements of upper molars as in *Scalopoides.*

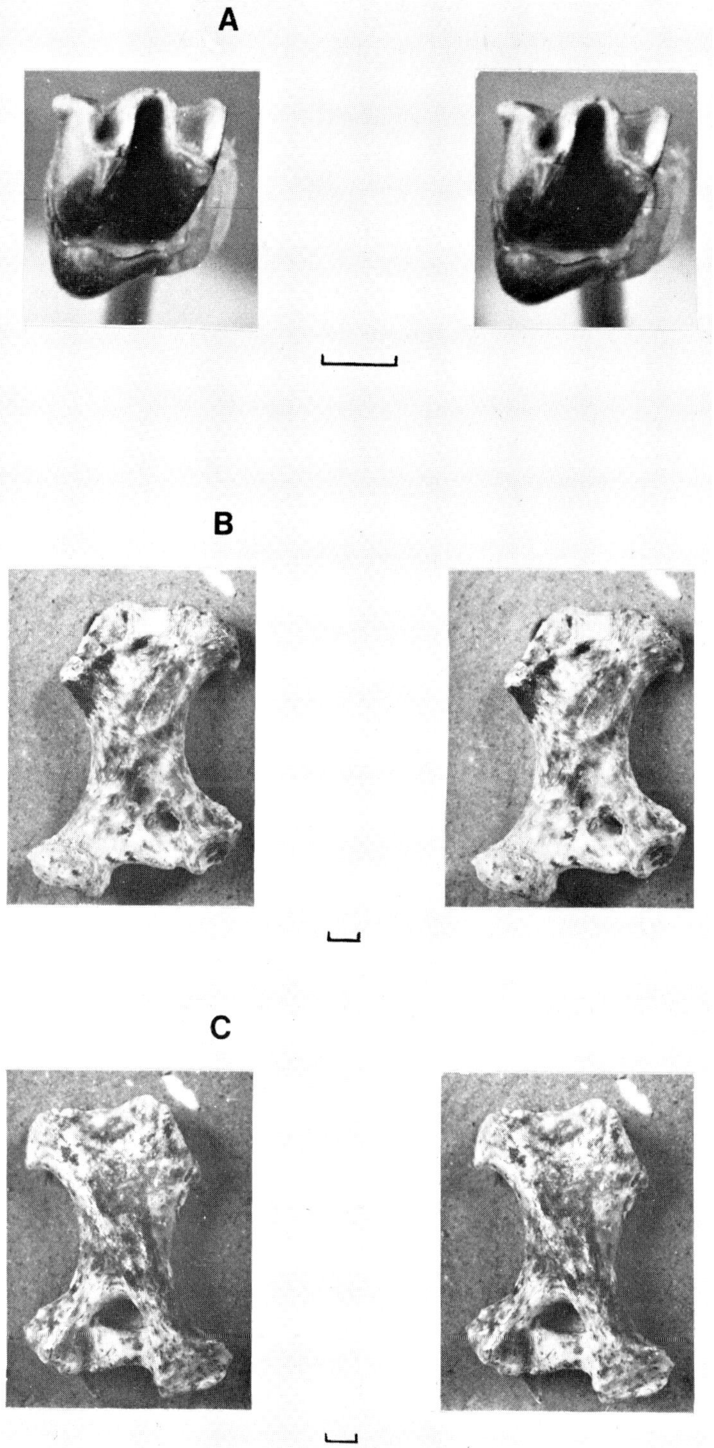

Fig. 7 *Scalopoides* sp., tooth and humerus. Scales 1 mm.

- A. LM^2, ROM 7656, anterior to left, buccal to top of page.
- B,C. R. humerus, NMC 8976, proximal to top of page.
- B. Anterior aspect.
- C. Posterior aspect.

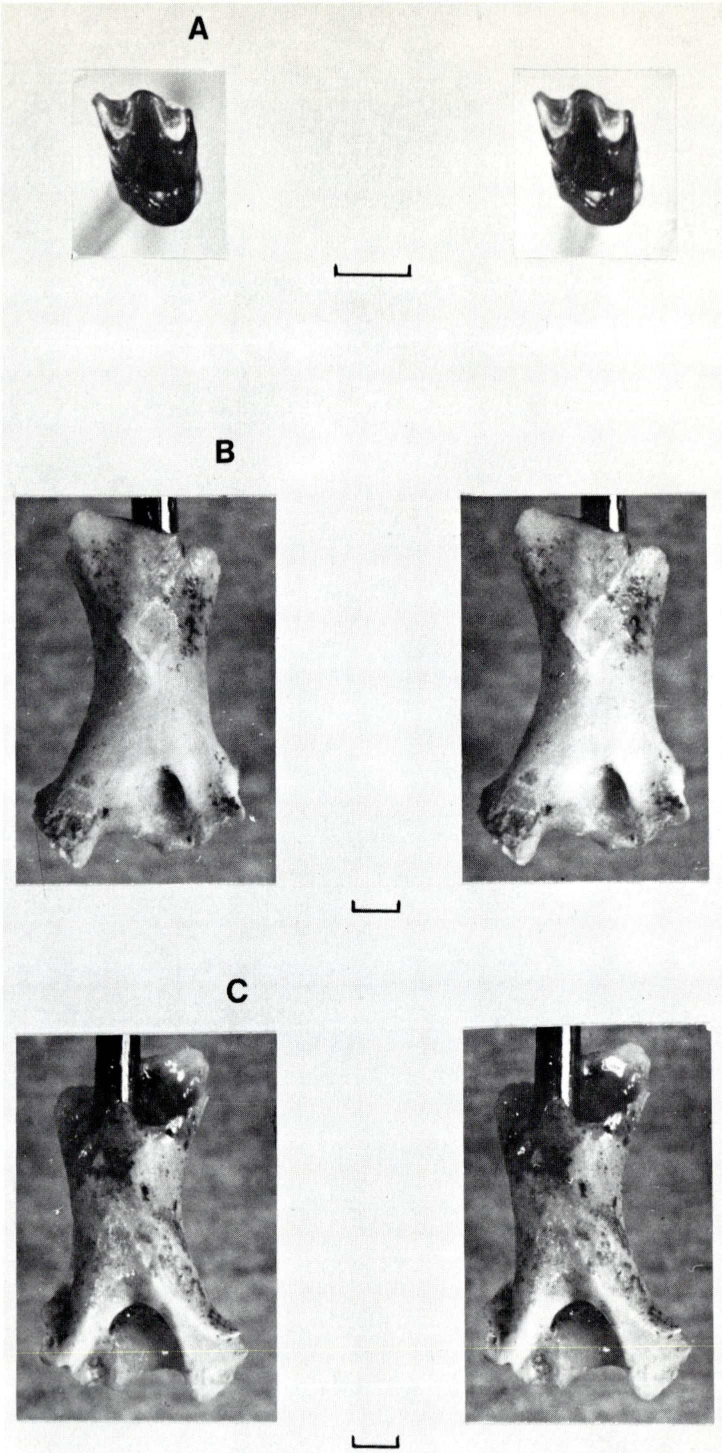

Fig. 8 Talpidae, gen. et sp. indet., tooth and humerus. Scales 1 mm.
A. RM^2, ROM 7667, anterior to right, buccal to top of page.
B,C. R. humerus, ROM 7668, proximal to top of page.
B. Anterior aspect.
C. Posterior aspect.

DISCUSSION

Insufficient dental material prevents detailed comparison with described specimens. Teeth are tentatively referred to the same species as the humerus. Future study may demonstrate that this material belongs to several species.

The humerus represents a weakly fossorial or aquatic mole (vide Hutchison, 1968). No illustrations that show close resemblance to ROM 7668 are known to me.

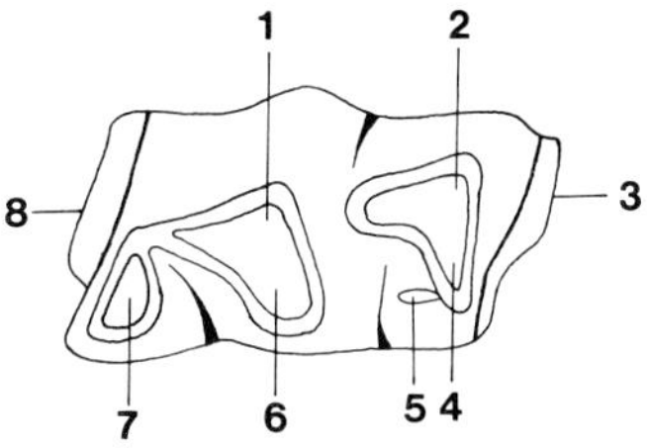

Fig. 9 Soricidae. RM_2, with terminology used in description of lower molars. Anterior to left, buccal to top of page. Not to scale.
1, protoconid; 2, hypoconid; 3, posterior cingulum; 4, entoconid; 5, entoconid ridge; 6, metaconid; 7, paraconid; 8, anterior cingulum.

Family Soricidae

?Trimylus sp.
Fig. 10A

REFERRED SPECIMEN. RM_1 (ROM 7621).

DESCRIPTION

M_1 moderately worn; loph connecting protoconid and metaconid lying at about 30° angle to midline of tooth; entoconid low; talonid basin shallow.

No cingula at enamel margins of crown; no entoconid crest present; no part of tooth fluorescing under ultraviolet light.

MEASUREMENTS

M_1, AP(1), 2.03; AP tri (1), 1.23; BL tri (1), 1.36.

DISCUSSION

The tooth lacks the strong entoconid crest of the Crocidurinae, Limnoecinae, and Soricinae, and failure of any part of the tooth to fluoresce indicates that it was not extensively pigmented as in the Blarininae. It is most likely that ROM 7621 represents a member of the Heterosoricinae from which pigmentation has disappeared through wear.

Among heterosoricines, the specimen shows some resemblance to *Trimylus dakotensis* (Repenning, 1967, p. 12). Sizes of the specimens are similar, and both show an "entoconid flush against but not confluent [with the hypoconid]". But there is no buccal cingulum on the Wood Mountain specimen. ROM 7621 is referred tentatively to *Trimylus* (=*Heterosorex* auct.) but differs from all described species in the absence of a buccal cingulum.

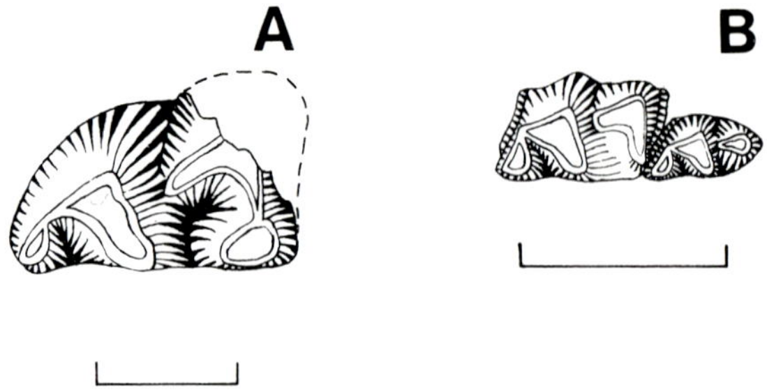

Fig. 10 ?*Trimylus* sp. and Heterosoricinae, gen. et sp. indet., lower molars in occlusal aspect. Anterior to left, buccal to top of page. Scales 1 mm.
A. ?*Trimylus* sp., RM_1, ROM 7621.
B. Heterosoricinae, gen. et sp. indet., RM_{2-3}, ROM 7623.

Heterosoricinae, gen. et sp. indet.
FIG. 10B

REFERRED SPECIMENS. L. mandibular fragment with LM_2 (ROM 7622), R. mandibular fragment with RM_{2-3} (ROM 7623).

DESCRIPTION

M_2 markedly rectangular; loph connecting protoconid and metaconid lying at 50°–60° angle to midline of tooth; entoconid low; no marked development of entoconid crest; low buccal cingulum, well defined, running from anterior end of paraconid to base of entoconid.

Features of M_3 trigonid identical with those of M_2; talonid abbreviated, one-cusped; posterior cingulum lacking.

Tips of cusps not fluorescing under ultraviolet light. Lower jaw shallow, delicate.

MEASUREMENTS

M_2, AP(2), OR 0.92–1.03; BL(2), OR 0.61–0.66. M_3, AP(1), 0.71; BL(1), 0.45.

DISCUSSION

Alluvisorex arcadentes (Hutchison, 1966, p. 18) shares the unusual feature of rectangularly shaped M_2 seen in the indeterminate material above. But the Wood Mountain teeth lack strong entoconid crests, and are not referable to *Alluvisorex* or any other member of the Soricinae. As with ?*Trimylus* sp. (*vide supra*), the lack of an entoconid crest indicates that the specimens represent a heterosoricine. Pigmentation, if present, must have covered only the tips of the cusps, as the little-worn tooth of ROM 7622 shows no pigmentation.

Order Carnivora
Family Felidae

Pseudaelurus cf. *P. aeluroides* Macdonald, 1954
Fig. 12

REFERRED SPECIMENS. LM_1 (ROM 7607), L. mandibular fragment with LM_1 (NMC 8964).

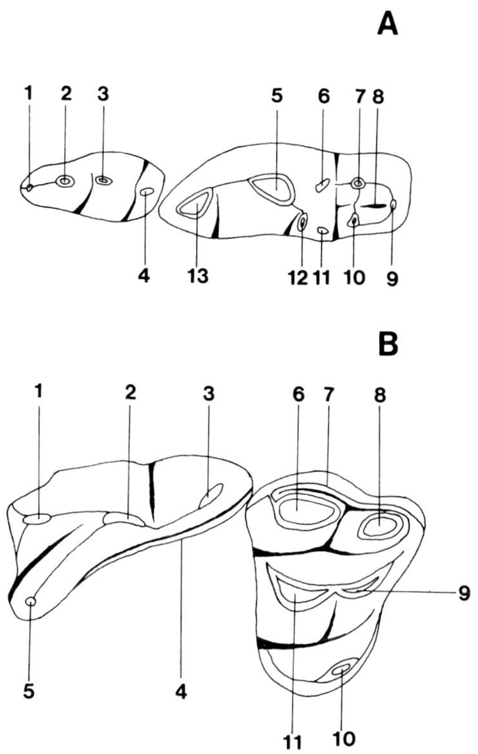

Fig. 11 Carnivora. Terminology used in description of cheek teeth. Anterior to left, buccal to top of page. Not to scale.

A. RP_4–M_1: 1, paraconid; 2, protoconid; 3, metaconid; 4, accessory cuspule on posterior shelf; 5, protoconid; 6, protostylid; 7, hypoconid; 8, talonid basin; 9, hypoconulid; 10, entoconid; 11, metaconulid; 12, metaconid; 13, paraconid.

B. LP^4–M^1: 1, parastyle; 2, paracone; 3, metacone; 4, lingual cingulum; 5, protocone; 6, paracone; 7, buccal cingulum; 8, metacone; 9, metastyle; 10, hypocone; 11, protocone.

DESCRIPTION

Lower carnassial (Fig. 12A) with paraconid and protoconid of about equal length; well-developed talonid; no metaconid. No sockets for M_2 in jaw; mandibular angle extending posterad as far as tip of condyle (Fig. 12B); masseteric crest prominent; two fossae for muscle attachment, separated by low, nearly horizontal ridge, in buccal face of angle, the more dorsal larger and deeper; single nutrient foramen on lingual face of angle (not shown).

MEASUREMENTS

M_1, AP(1), 15.7; BL(1), 6.8.

DISCUSSION

The surface of NMC 8964 is pitted and waterworn, but nevertheless LM_1 agrees well in structure and size with that of the holotype of *Pseudaelurus aeluroides*,

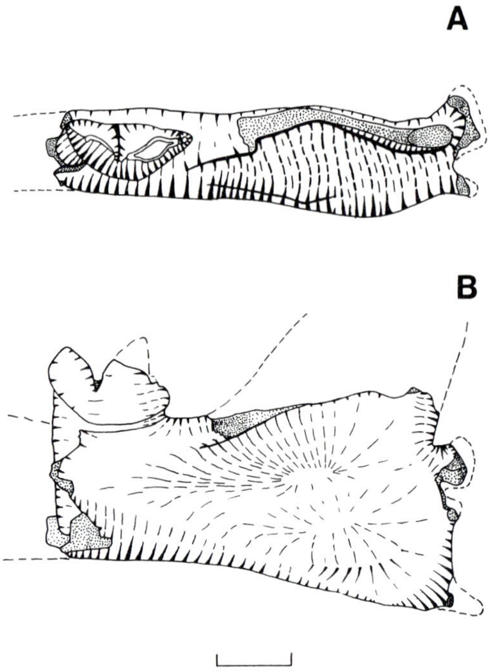

Fig. 12 *Pseudaelurus* cf. *P. aeluroides*, L. mandibular fragment with LM_1, NMC 8964. Anterior to left. Scale 1 cm.
A. Occlusal aspect, buccal to bottom of page.
B. Lateral aspect, occlusal to top of page.

AMNH 22398 from the Lower Snake Creek beds; carnassials of NMC 8964 and AMNH 22398 are 15.7 and 15.6 in anteroposterior, and 6.8 and 7 (est.) in buccolingual diameter, respectively (measurements of AMNH 22398 from Macdonald, 1954, pp. 67–68). There are two nutrient foramina on the lingual face of the mandibular angle in the type, but NMC 8964 has only one. M_1 is nearer the posterior end of the angle in the type than in NMC 8964, because the type is a young individual in which M_1 is not fully erupted. A toothless jaw (AMNH 22404), from the same beds as the holotype, is from an older individual and agrees with NMC 8964 in all observable details.

Family Mustelidae

Sthenictis cf. _S. dolichops_ Matthew, 1924
Fig. 13

REFERRED SPECIMENS. LP_2 (ROM 7478), L. mandibular fragment with LP_{2-4}, LM_1 (NMC 10023), L. mandibular fragment with LP_4 (GSC XXII), LM^2 (SMNH P552).

DESCRIPTION
Of premolars, only P_4 showing "stepped" outline with metaconid strongly developed and showing clearly in lateral view; posterior shelves of premolars laterally expanded more in P_4 than in P_3, and more in P_3 than in P_2.

Lower carnassial badly broken, lacking protoconid, which was probably about one and a half times as high as paraconid, and about equal in anteroposterior length; distinct, prominent metaconid, about as high as paraconid, buccal and posterior to protoconid; talonid high and as long as paraconid or protoconid, apparently consisting of only high trenchant hypoconid with low loph running distally from it; entoconid presumably absent.

Ventral border of lower jaw (Fig. 13B) strongly convex; teeth mentioned above, alveoli for P_1 and M_2 present; two mental foramina, smaller beneath centre of P_3, larger beneath centre of P_2.

DISCUSSION

Lower jaw NMC 10023 is larger than the holotypic jaw of *Sthenictis bellus*, AMNH 20501, from the Sheep Creek Fauna, and the ramus is deeper dorsoventrally. No foramen is present on the lingual side of the mandibular angle in NMC 10023, and mental foramina are more prominent than in *S. bellus*.

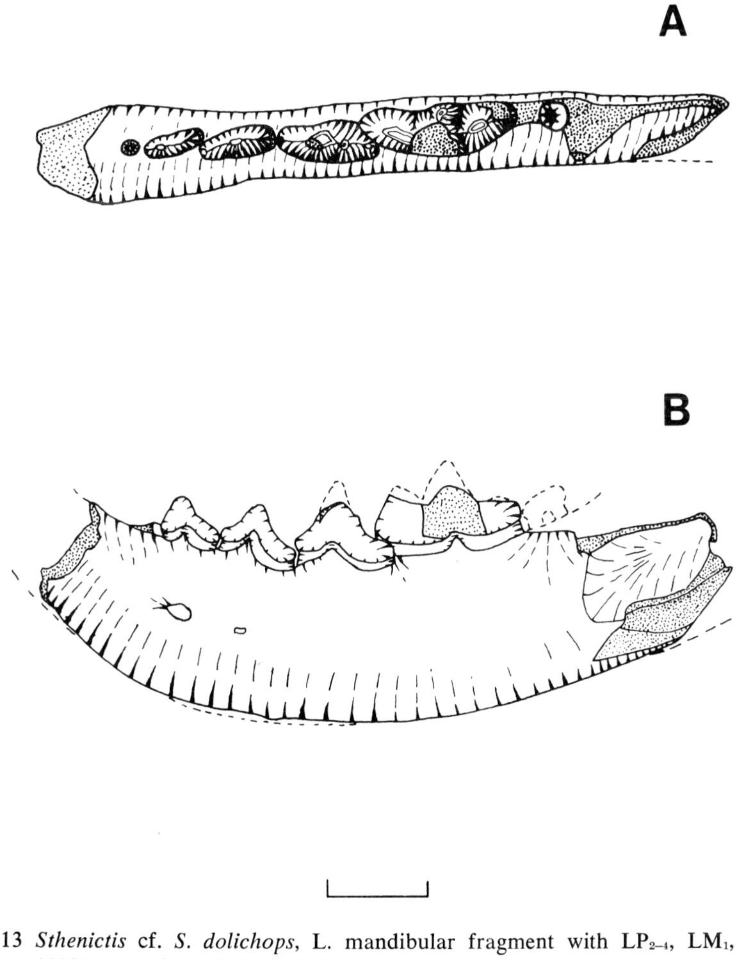

Fig. 13 *Sthenictis* cf. *S. dolichops*, L. mandibular fragment with LP_{2-4}, LM_1, NMC 10023. Anterior to left. Scale 1 cm.
A. Occlusal aspect, buccal to bottom of page.
B. Lateral aspect, occlusal to top of page.

The holotypic jaw of *Sthenictis dolichops*, AMNH 18264, from the Lower Snake Creek Fauna, has teeth of similar sizes to those of NMC 10023 (Table 1) and mental foramina in about the same positions. P_2 and P_3 are less crowded in AMNH 18264, the premolar series is more open, and the ramus is deeper dorsoventrally. NMC 10023 is similar in nearly all characteristics to the holotype of *S. dolichops*.

Table 1. Measurements of teeth of *Sthenictis* cf. *S. dolichops* compared with those of the holotype of *S. dolichops* (estimated from Matthew, 1924, fig. 34).

Tooth	Dimension	n	OR	Holotype of *S. dolichops*, AMNH 18264
P_2	AP	2	6.4–6.7	6.5
	BL	2	2.9–2.9	3.5
P_3	AP	1	8.0	9
	BL	1	3.3	4
P_4	AP	2	9.0–9.8	11
	BL	2	4.4–5.2	5
M_1	AP	1	15.0	16
	BL	1	5.3	6.5
M^2	AP	1	8.5	
	BL	1	9.7	

Miomustela cf. *M. madisonae* (Douglass), 1929

Fig. 14

REFERRED SPECIMEN. R. mandibular fragment with RP_4, RM_1, alveolus for RM_2 (SMNH P558).

DESCRIPTION

Teeth and jaw surface badly broken and waterworn.

P_4 showing definite anterior shelf, incipient paraconid; posterior shelf prominent, lacking accessory cuspules.

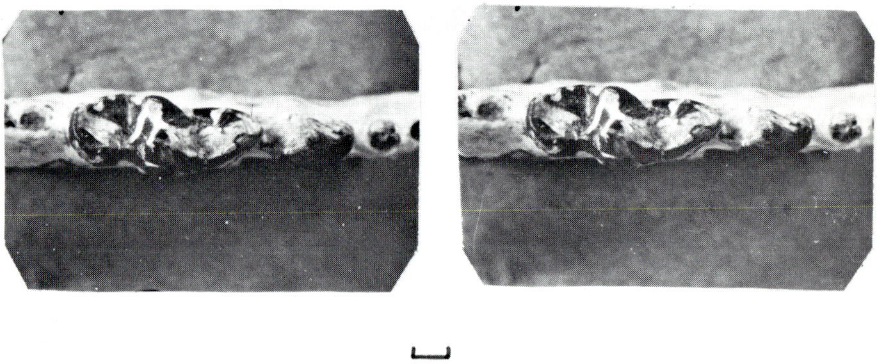

Fig. 14 *Miomustela* cf. *M. madisonae*, R. mandibular fragment with RP_4, RM_1, SMNH P558. Occlusal aspect. Anterior to right, buccal to bottom of page. Scale 1 mm.

Lower carnassial with paraconid and talonid of about equal length, both slightly shorter than protoconid; large metaconid as high as paraconid; talonid basin opening lingually through wide, deep trough; poor condition of tooth precludes determination of relative sizes of hypoconid and entoconid, although both probably present; hypoconid separated from protoconid by notch.

Horizontal ramus of lower jaw shallow dorsoventrally; ventral border shallowly convex; mental foramen beneath anterior root of P_4.

DISCUSSION

The size of the jaw and the size and proportions of M_1 (Table 2) closely approach the diagnosis of *Miomustela madisonae* (Hall, 1930, p. 152) except for the location of the mental foramen and possibly the length of P_4.

Table 2. Measurements of teeth of *Miomustela* cf. *M. madisonae* compared with those of the holotype of *M. madisonae* (from Hall, 1930, p. 153).

Tooth	Dimension		SMNH P558	Holotype of *M. madisonae*, CM 848
P_4	AP		2.7	3.4
M_1	AP		5.3	5.1
	BL	tri	2.2	2.2
	BL	tal	1.8	1.8

Plionictis cf. *P. ogygia* (Matthew), 1901

Fig. 15

REFERRED SPECIMEN. R. mandibular fragment with RdP_4, RM_1, alveoli for RP_3 and RM_2 (NMC 8967).

DESCRIPTION

Deciduous fourth lower premolar smaller than but similar to permanent carnassial; large, prominent protoconid twice as high as paraconid; metaconid distinct, low; talonid low, basined; protoconid-hypoconid ridge stronger than in permanent carnassial; strong hypoconid, cingular entoconid as in permanent carnassial.

Permanent lower carnassial showing stronger paraconid than deciduous equivalent; protoconid not fully twice height of paraconid; small, low metaconid, well separated from protoconid, posterior and lingual to protoconid; protostylid small; large hypoconid on talonid; entoconid high, arcuate ridge at anterior enamel margin; hypoconid weakly connected to protoconid; entoconid and metaconid separated by wide, deep valley; talonid trenchant in function, separation between entoconid and metaconid leaving lingual opening of talonid basin very wide and deep.

Lower jaw shallow vertically; ventral border nearly horizontal beneath tooth row, possibly deepening near poorly preserved angle; large masseteric fossa occupying most of angle; large mental foramen located beneath posterior socket for P_3, near midlength on root; single socket for the second lower molar preserved.

MEASUREMENTS

dP_4, AP(1), 4.7; BL(1), 2.2. M_1, AP(1), 7.3; BL(1), 3.2.

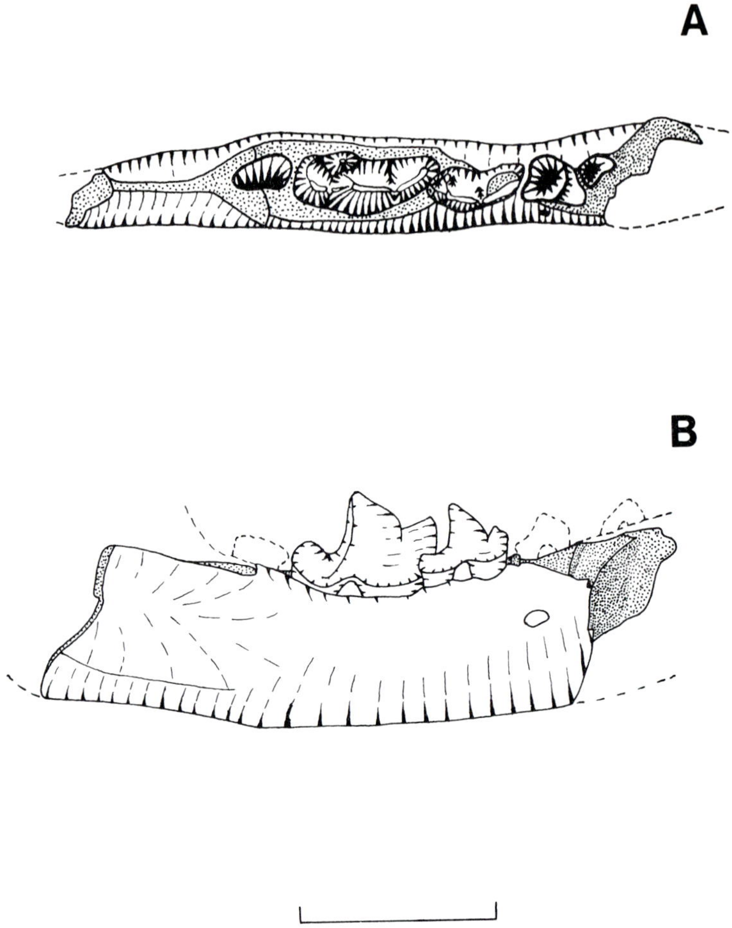

Fig. 15 *Plionictis* cf. *P. ogygia*, R. mandibular fragment with RdP_4, RM_1, NMC 8967. Anterior to right. Scale 1 cm.
A. Occlusal aspect, buccal to bottom of page.
B. Lateral aspect, occlusal to top of page.

DISCUSSION

The specimen, NMC 8967, was compared with the holotype of *Plionictis ogygia*, AMNH 9042, from the Pawnee Creek Beds: the specimens agree well in size and in depth and shape of the jaw, M_1 7.3 and 8 in anteroposterior, and 3.2 and 3.3 in buccolingual diameter respectively (measurements of AMNH 9042 estimated from Matthew, 1901, fig. 9). M_1 from Wood Mountain is nearly unworn and the AMNH holotype is heavily worn, but these teeth show close resemblance in all characters.

Mustelidae, gen. et sp. indet.

Fig. 16

REFERRED SPECIMEN. L. mandibular fragment with alveoli for LP_4, LM_{1-2} (SMNH 7987).

DESCRIPTION

Horizontal ramus shallow, moderately curved; large fossa for insertion of masseter at angle; position of mental foramen not preserved; P_4 crowding into M_1 as in *Sthenictis* and *Plionictis*; septa between sockets for single teeth well developed, reaching to tops of outer edges of alveoli.

MEASUREMENTS

Ramus, DV beneath M_1(1), 10.7; BL beneath M_1(1), 5.3; AP of M_1 alveolus (1), 9.1; BL of M_1 posterior socket (1), 2.8.

DISCUSSION

In Fig. 16, I have made a rough restoration of the teeth. In my interpretation, P_4 and M_1 were two-rooted, and M_2 was one-rooted.

Among Wood Mountain carnivores, this jaw is too small to represent *Sthenictis dolichops* or *Leptocyon vafer*, and is too large to be *Bassariscus, Miomustela*, or *Plionictis*. The mandible is from an adult individual, as the septa between sockets are complete, showing that M_1 was fully erupted, and the alveolus for M_2 is large and open, indicating that the second molar also had erupted. Curvature of the ramus and possible shape of M_1 resemble Wood Mountain specimens referred to *Sthenictis*, and the specimen probably represents a medium-sized mustelid.

Less probable is that the fragment represents a small canid. Macdonald (1948, fig. 2) figured a canid of equivalent size from the Clarendonian Black Hawk Ranch Fauna but made no generic assignment, stating only that the taxon could be related to *Urocyon*. The length of the alveolus for M_1 for Macdonald's specimen was 9.3 mm, a figure close to that obtained for the Wood Mountain ramal fragment. These specimens definitely belong to different taxa, however, because the alveolus for M_1 is shaped differently, and Macdonald's specimen has two-rooted M_2 and shows no crowding of P_4 and M_1.

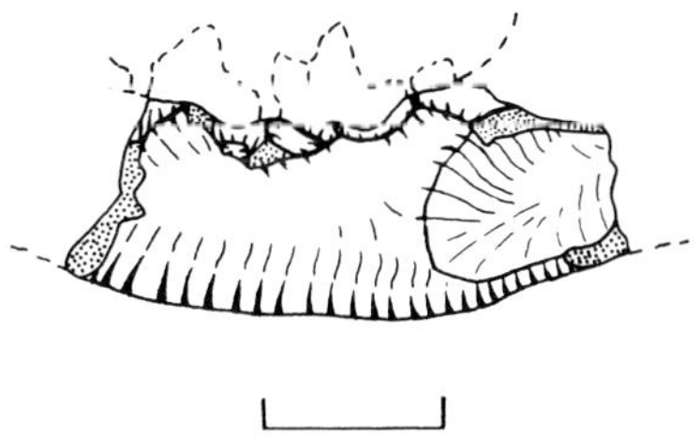

Fig. 16 Mustelidae, gen. et sp. indet., L. mandibular fragment, SMNH 7987. Lateral aspect. Anterior to left, dorsal to top of page. Scale 1 cm.

Family Canidae

Tomarctus hippophagus (Matthew and Cook), 1909

Fig. 17

REFERRED SPECIMENS. ?LC_1 (NMC 9438), LP_2 (SMNH P237), LP_3 (SMNH P237), L. mandibular fragment with LP_{2-4}, LM_{1-2} (NMC 9077), L. mandibular fragment with LP_4, broken LP_3 and LM_1 (SMNH P550), R. mandibular fragment with RP_{2-4}, broken RM_1 (SMNH 8494), R. mandibular fragment with sockets for RP $_{1-4}$, RM_1 (ROM 7477), LM_2 (ROM 7483), RM_2 (NMC 9438), L. mandibular fragment with posterior socket for LM_1, full sockets for LM_2 (NMC 8974) RP^4 (SMNH P123; ROM 7476), LM^1 (SMNH P552), RM^1 (SMNH 7993), R. maxillary fragment with alveoli for RP^4 and RM^1 (NMC 9074), R. maxillary fragment with alveolus for RM^1 (NMC 9439).

DESCRIPTION

P_2 to P_4 (Fig. 17A, B) not crowded together, evenly spaced; progressive increase from P_2 to P_4 in overall size, proportional buccolingual width, buccolingual expansion of posterior shelf, size of metaconid, size of accessory cuspule on posterior shelf; last three lower premolars with prominent metaconids, "stepped" outline in lateral view; in several specimens of P_4, including NMC 9077, faint incipient paraconid appearing at anterior enamel margin; P_1 one-rooted, not represented in collections from Wood Mountain Formation.

In lower carnassial of NMC 9077, angle at anterobuccal corner of crown relatively sharp; high, prominent protoconid occupying much of trigonid; paraconid high; metaconid low, small, well separated from protoconid, posterolingual to it; protostylid and metaconulid small; hypoconid, entoconid large, connected by transverse ridge, with valleys running anterad and posterad in midline of tooth.

In M_2, protoconid and metaconid more prominent than paraconid; metaconid posterolingual to protoconid; cingulum occupying anterobuccal enamel margin; talonid strongly basined; large hypoconid and entoconid connected posteriorly by arcuate ridge, and sending ridges forward to bases of protoconid and metaconid; trigonid and talonid of approximately equal length.

Mandibular ramus moderately deep dorsoventrally; ventral border running horizontally beneath tooth row, curving dorsad beneath angle; mental process small; mental foramina beneath centres of P_2 and P_3.

P^4 (Fig. 17C) with small but distinct protocone, lacking parastyle; paracone large, prominent, connected to high, anteroposteriorly expanded metacone by

Table 3. Measurements of teeth of Wood Mountain *Tomarctus hippophagus* compared with those of holotype, *T. hippophagus* (from Matthew and Cook, 1909, p. 375).

Tooth	Dimension	n	Mean	OR	Holotype of *T. hippophagus*, AMNH 13836
C_1	AP	1		8.2	
	BL	1		5.4	
P_2	AP	3	7.1	6.5–7.7	8.5
	BL	3	3.8	3.3–4.4	5.0
P_3	AP	3	8.7	8.4–9.1	9.4
	BL	3	4.4	3.7–5.2	6.0
P_4	AP	2		10.1–10.4	12.0
	BL	2		5.8–6.3	6.8
M_1	AP	1		20.0	19.7
	BL	1		8.1	9.0
M_2	AP	3	10.9	10.1–12.4	10.3
	BL	3	6.8	5.8–7.9	7.2
P^4	AP	2		19.0–21.6	
	BL	1		10.1	

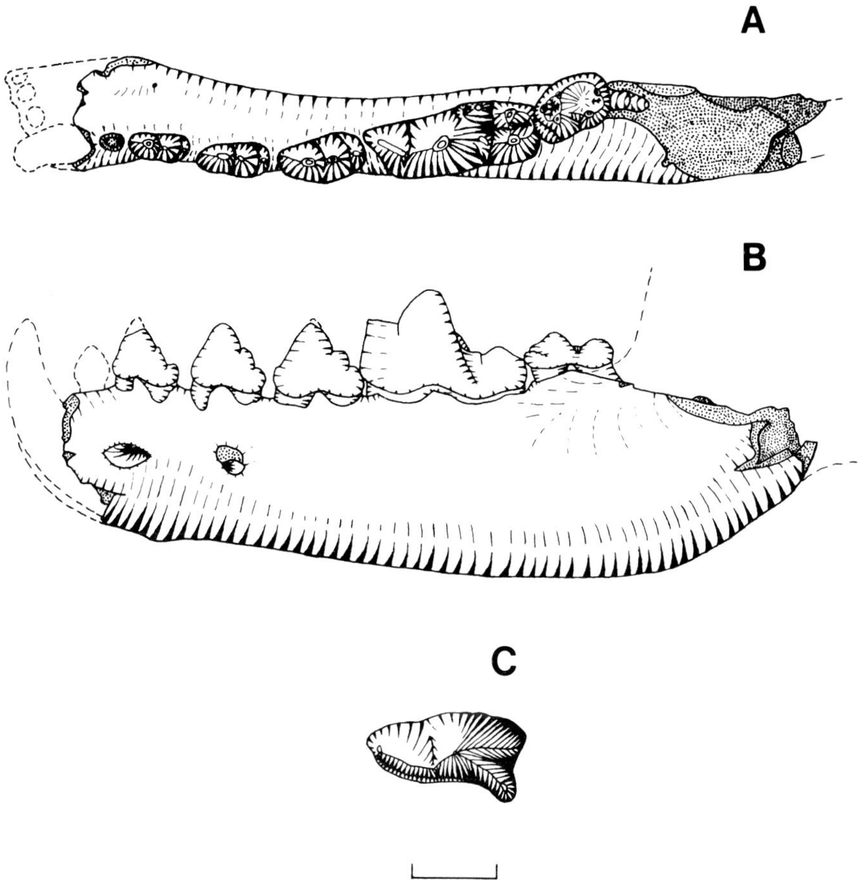

Fig. 17 *Tomarctus hippophagus*, mandibular fragment and P^4. Scale 1 cm.
A,B. L. mandibular fragment with LP_{2-4}, LM_{1-2}, NMC 9077, anterior to left.
A. Occlusal aspect, buccal to bottom of page.
B. Lateral aspect, occlusal to top of page.
C. RP^4, SMNH P123, anterior to right, buccal to top of page.

low ridge; lingual cingulum traversing enamel margin from anterior end of metacone.

DISCUSSION

In size and shape of teeth (Table 3) and jaw, pattern of lower teeth, and positions of mental foramina, NMC 9077 (Fig. 17A, B) matches nearly perfectly AMNH 13836, the holotype of *Tomarctus hippophagus*. Particularly important is the transverse expansion of posterior shelves of P_2 to P_4, similarly pronounced in NMC 9077 and AMNH 13836. Protostylids and metaconulids are found in the lower carnassials of both specimens; however, these cuspules are variably present in many canids.

The two upper carnassials compare closely with material referred to *T. hippophagus* and to other species of *Tomarctus* in the collections of the American Museum of Natural History. They are of appropriate size for reference to *T. hippophagus*, but bear no special features diagnostic of any particular species of *Tomarctus*.

Aelurodon cf. *A. saevus* (Leidy), 1858
Fig. 18

REFERRED SPECIMENS. R. mandibular fragment with roots of RC_1, RP_2, anterior root of RP_3 (SMNH P239), LP_4 (SMNH P391), L. mandibular fragment with LP_4, alveoli for LM_{1-2} (SMNH P563), RM_1 (SMNH P123), R. mandibular fragment with alveoli for RM_{2-3} (ROM 7484), LP^4 (SMNH P379, P391; ROM 7608), RP^4 (ROM 7475), RM^1 (SMNH P391).

DESCRIPTION

P_4 (not shown) differing from corresponding tooth of *Tomarctus hippophagus* in its greater size, proportionally greater width; paraconid and metaconid prominent, "stepped" pattern well developed; strong accessory cuspule on posterior shelf; posterior shelf not expanded transversely as much as in *Tomarctus hippophagus.*

Lower carnassial (Fig. 18A) robust; protoconid barely higher than paraconid; well-defined metaconid much lower than paraconid; protostylid small; anterobuccal enamel margin forming cingulum beneath paraconid.

Talonid of lower carnassial low; hypoconid and entoconid prominent, connected by strong transverse ridge; arcuate ridge, not bearing hypoconulid, connecting hypoconid and entoconid posteriorly; accessory cuspule just anterior to entoconid.

Mandibular ramus (not shown) resembling that of *Tomarctus hippophagus* in dorsoventral depth (Fig. 17A, B), but ventral border more strongly curved; mental foramen beneath centre of P_3.

Upper carnassial (Fig. 18B) with large parastyle; paracone just higher than metacone; protocone high, broad; low lingual cingulum running from base of paracone to posterior tip of metacone.

M^1 (not shown) large, but without diagnostic characters; typical of medium-sized canids.

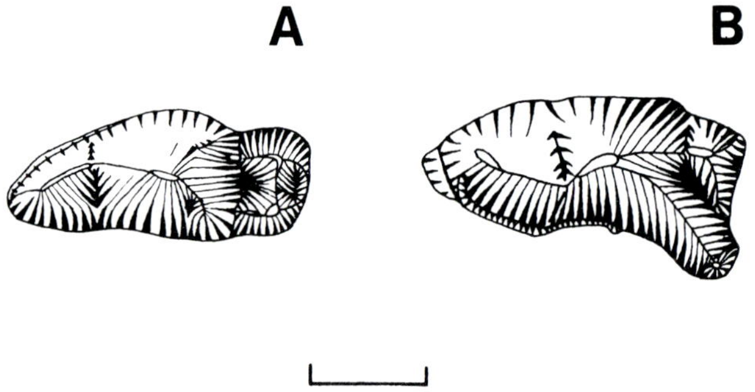

Fig. 18 *Aelurodon* cf. *A. saevus*, teeth in occlusal aspect. Buccal to top of page. Scale 1 cm.
A. RM_1, SMNH P123, anterior to left.
B. RP^4, ROM 7475, anterior to right.

DISCUSSION

These specimens fit applicable parts of the diagnosis of *Aelurodon* (VanderHoof and Gregory, 1940, p. 144): "A well-developed parastyle on P^4; the anterior premolars of both upper and lower jaws stepped behind ...; an anterior basal cusp on the premolars may be present or absent; the dentition is heavy and the premolars are not reduced greatly in size,and never abruptly reduced."

Wood Mountain specimens have not been compared with the holotype of *Aelurodon saevus* (VanderHoof and Gregory, 1940, p. 145), but the lower carnassial resembles the type lower carnassial in size (Table 4) and visible characters, although the holotype is more worn than SMNH P123. Good agreement is shown with AMNH 8306, referred to *A. saevus*, from the "*Procamelus*" beds in Hitchcock County, Nebraska. The Wood Mountain material is therefore referred to *Aelurodon* cf. *A. saevus*.

Table 4. Measurements of teeth of *Aelurodon* cf. *A. saevus* compared with those of holotype of *A. saevus* (estimated from VanderHoof and Gregory, 1940, fig. 2).

Tooth	Dimension	n	OR	Holotype of *A. saevus*, USNM 127
P_4	AP	2	12.0–15.1	14.5
	BL	2	6.2–8.3	
M_1	AP	1	25.4	27
	BL	1	10.2	11
P^4	AP	2	25.8–27.8	
	BL	1	13.6	
M^1	AP	1	17.1	
	BL	1	19.5	

Leptocyon cf. *L. vafer* (Leidy), 1858
Fig. 19

REFERRED SPECIMENS. LM_1 (SMNH 7989, 7993), RM_1 (SMNH 7993; NMC 9438).

DESCRIPTION

Lower carnassials with protoconids longer than and almost as high as paraconids; metaconids almost entirely separated from protoconids, not as high as paraconids; talonids low, basined, with small entoconids, slightly larger hypoconids; low cristae reaching forward from entoconids and hypoconids; talonids about as long anteroposteriorly as protoconids.

MEASUREMENTS

M_1, AP(3), mean 11.3, OR 10.3–12.3; BL(3), mean 4.5, OR 3.9–5.0.

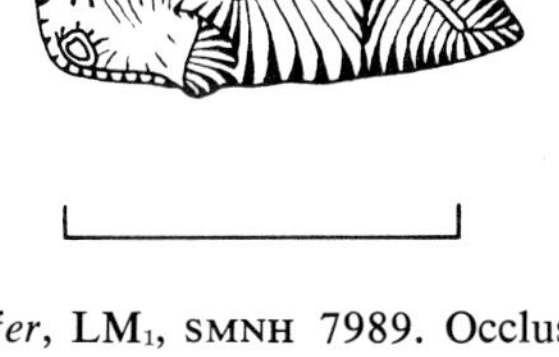

Fig. 19 *Leptocyon* cf. *L. vafer*, LM_1, SMNH 7989. Occlusal aspect. Anterior to right, buccal to top of page. Scale 1 cm.

DISCUSSION

SMNH 7989 was compared to AMNH 8546, 17201, 18356, and 18265, lower carnassials of *Leptocyon vafer* from the Lower Snake Creek Beds. These specimens resemble SMNH 7989 in all visible details, including proportions and characteristics of the talonids. AMNH 8548 is slightly smaller but shows identical dental characters. AMNH 20053 has a smaller metaconid, but its talonid resembles that of SMNH 7989. SMNH 7989 fits well within *Leptocyon vafer*, and the other three Wood Mountain specimens, identified after the above comparison was made, agree well with SMNH 7989 and with each other.

Family Amphicyonidae

Amphicyon sp.

Figs. 20, 21

REFERRED SPECIMENS. Fragment of skull with LP^2, broken LP^3, LP^4, RP^{2-4}, LM^1, RM^1, alveoli for LI^{1-3}, RI^{1-3}, LC^1, RC^1, LP^1, RP^1, LM^{2-3} (ROM 7698), LM^1 (NMC 9385).

DESCRIPTION

Skull fragment and isolated M^1 present in collections. Dental formula I^3, C^1, P^4, M^3; incisor three largest, incisor one smallest; large, oval canine separated from third upper incisor by short space.

Premolars well separated from each other; P^1 one-rooted; P^1 not preserved on either side of palate, but apparently small, peglike.

P^2 two-rooted; small anterolingual protocone, posterior metacone connected by strong ridges to large central paracone; P^3 similar in structure to P^2, slightly larger.

Upper carnassial, P^4, proportionally much larger than P^3; prominent parastyle connected by strong ridge to large central paracone; protocone large, isolated, lying in position typical for amphicyonids, posterolingual to parastyle rather than anterolingual; long, high metacone directed toward protocone, separated from paracone by deep notch; no cingula present on P^4, but enamel margin raised slightly posterolingual to metacone; P^4 three-rooted, as M^1 and M^2, with roots under paracone, metacone, and protocone.

M^1 broadest tooth (transversely) in upper tooth row; paracone higher, slightly larger than metacone; buccal cingulum weak; low, bulbous, oval protocone separated from hypocone by deep valley; protocone connected anteriorly with small protoconule, posteriorly with low, ridge-like metaconule; shallow valleys separating protoconule and metaconule from paracone and metacone; oval basin occupying centre of crown; hypocone occupying posterolingual portion of tooth, nearly as high as protocone.

Alveoli for M^2 and M^3 preserved on left side of palate, but teeth themselves not recovered; two buccal roots of M^2 nearly fused, inserted in yoked sockets; M^2, estimating from distance between lingual, buccal sockets, nearly as wide buccolingually as M^1; M^3 small, single-rooted.

Slit-like, anteroposteriorly directed incisive foramen lying behind I^1 on each side; premaxillary-maxillary suture running straight mediad from posterolingual edge of alveolus for canine; anterior borders of internal nares apparently lying posterior to the posterior end of M^3, as in *Amphicyon sinapius* (Matthew, 1924, fig. 22).

In basicranial region, (Figs. 20B, 21B), glenoid fossa wide transversely and dorsoventrally; external auditory meatus large, trench-like, as in other amphicyonids, large process of petrosal blocking external auditory meatus ventrally.

Palatal portion occupying less than half length of skull (Figs. 20A, 21A); nasal opening extending posteriorly as far as anterior end of C^1, in line perpendicular to surface of palate; anterior portions of nasal bones broken away; infraorbital foramen oval, facing directly anteriorly.

Temporal table flat dorsally; stout jugals flaring posterolaterad, slightly dorsad.

Sagittal crest very high, long, stout; broadening laterally at posterior end to meet strong lambdoid crest; lambdoid crest enclosing area of posterior end of skull which is high dorsoventrally, comparatively narrow laterally, bisected by smaller median crest; on left side (side shown) knobs lying ventral to posterior portion of sagittal crest, anterior to dorsal portion of lambdoid crest; no knobs on right side (not shown).

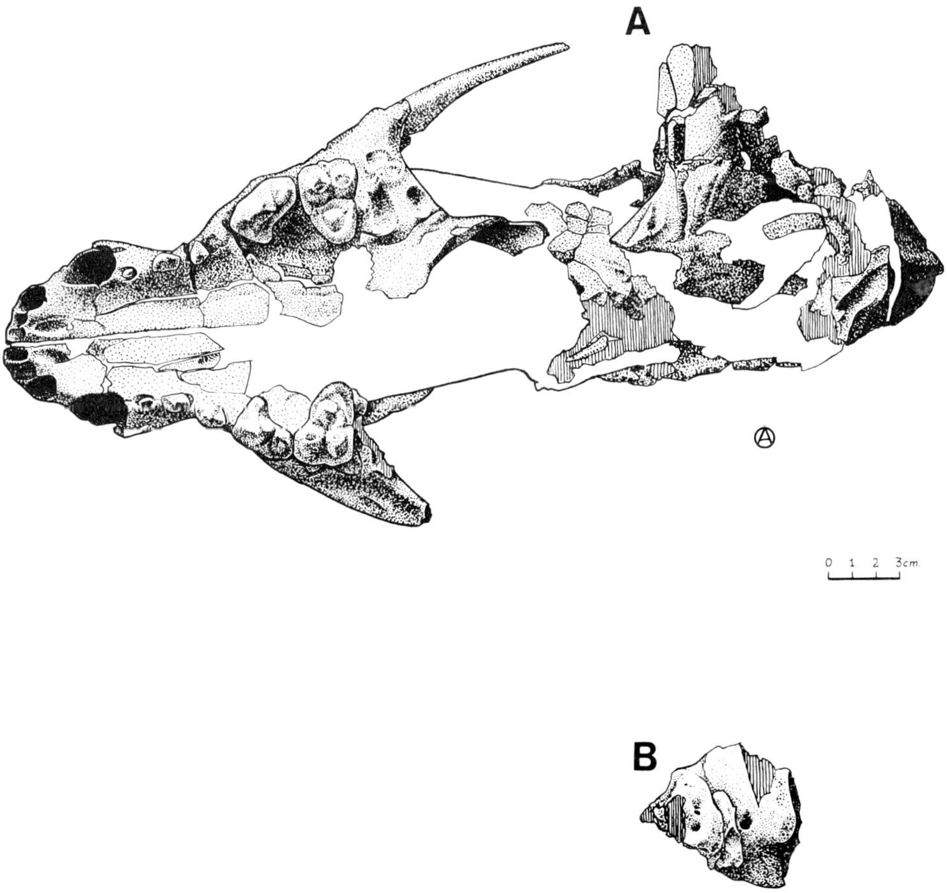

Fig. 20 *Amphicyon* sp., skull fragment, ROM 7698. Palatal aspects. Scale 3 cm.
A. Skull fragment, left to top of page, anterior to left.
B. Right ear region, left to bottom of page, anterior to right.

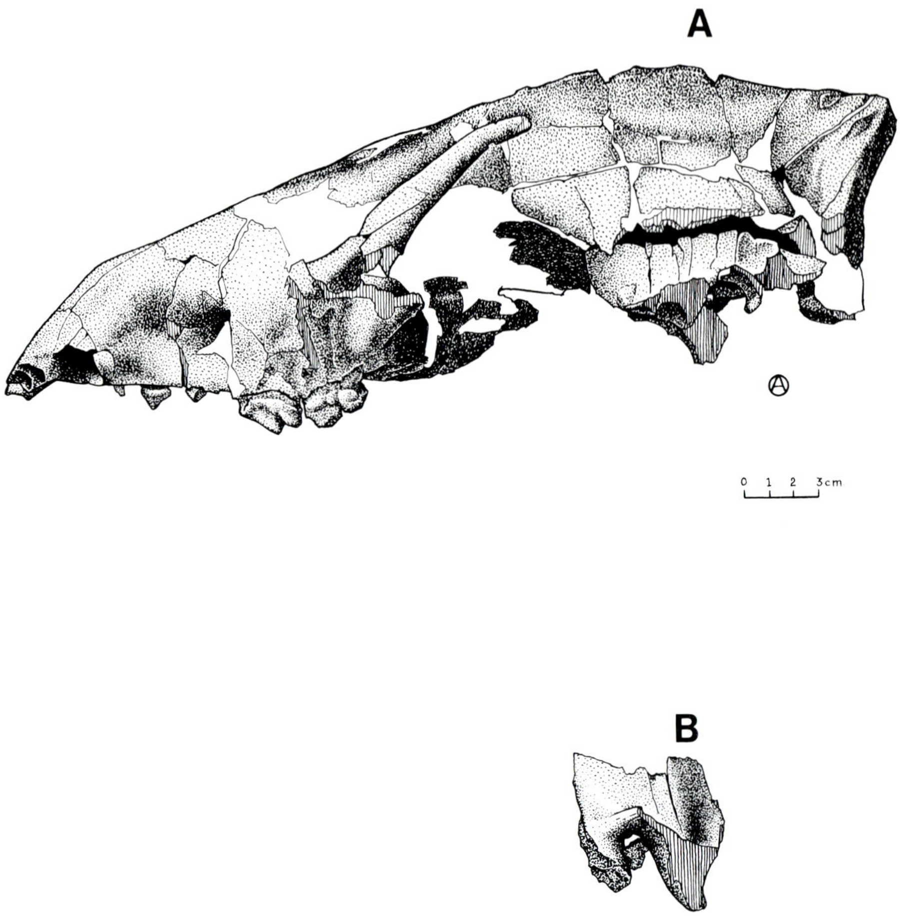

Fig. 21 *Amphicyon* sp., skull fragment, ROM 7698. Lateral aspects. Dorsal to top of page. Scale 3 cm.
A. Skull fragment, anterior to left.
B. Right ear region, anterior to right.

MEASUREMENTS

P^2, AP(2), OR 12.0–12.1; BL(2), OR 17.5–17.6. P^3, AP(1), 13.2; BL(1), 18.7. P^4, AP(2), OR 25.8–27.2; BL(2), OR 17.2–18.0. M^1, AP(3), mean 21.9, OR 20.7–23.1; BL(3), mean 30.7, OR 29.3–32.4.

DISCUSSION

The skull fragment described above shows differences from each of the three described North American genera of Upper Miocene amphicyonids. Although there are three upper molars and M^1 and M^2 are subequal, as in *Amphicyon*, P^3 is a small tooth relative to P^4 in the Wood Mountain specimen, whereas in described *Amphicyon* it is relatively large. In *Pliocyon*, M^1 and M^2 are subequal and P^3 is relatively small, but there are only two upper molars. In *Ischyrocyon* there can be three upper molars, but M^2 is only two-thirds as wide as M^1 (Webb, 1969, p. 46). The Wood Mountain specimen shows fewer differences from *Amphicyon* than from *Pliocyon* or *Ischyrocyon* and is referred to that genus.

Family Procyonidae

***Bassariscus* cf. *B. parvus* Hall, 1927**
Figs. 22, 23

REFERRED SPECIMENS. L. mandibular fragment with LP_4 (GSC XXIV), L. mandibular fragment with LP_4 and LM_1 (SMNH P123), LM_2 (ROM 7482), RM_2 (ROM 7480, 7481), R. mandibular fragment with RM_2 (ROM 7479).

DESCRIPTION

P_4 with high protoconid (Fig. 22A, B); posterior end showing "stepped" profile; large posterior shelf; strong ridge connecting protoconid to prominent metaconid, leaving posterior surface of protoconid flat.

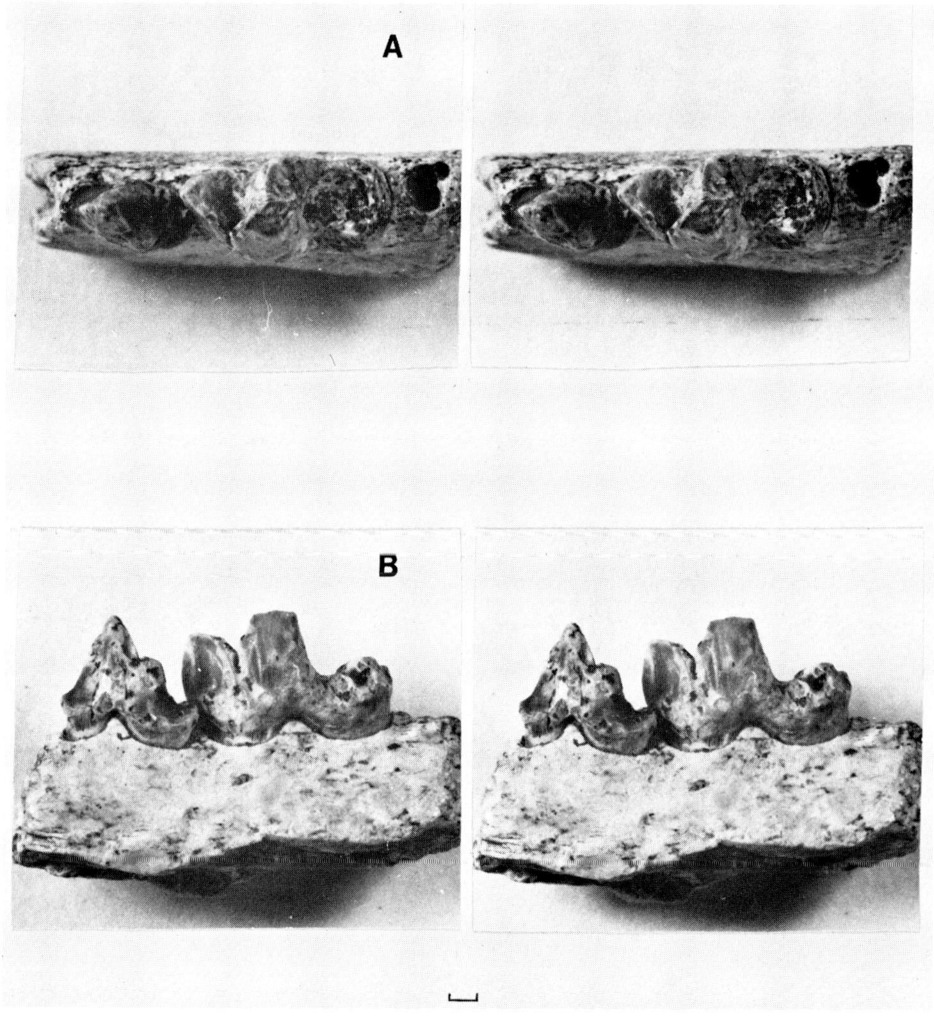

Fig. 22 *Bassariscus* cf. *B. parvus*, L. mandibular fragment with LP_4, LM_1, SMNH P123. Anterior to left. Scale 1 mm.
A. Occlusal aspect, buccal to bottom of page.
B. Lateral aspect, occlusal to top of page.

Lower carnassial protoconid, talonid of about equal length, each longer than paraconid; paraconid, large, distinct metaconid of equal height, nearly as high as protoconid; posterior face of metaconid barely posterior to protoconid; talonid basinal, lower on lingual side; small hypoconid and entoconid forming crest closing off posterior and lingual sides of talonid; shallow valleys between hypoconid and protoconid, and between entoconid and metaconid.

Trigonid, talonid of M_2 subequal (Fig. 23); low cuspule lingual to paraconid creating trigonid basin; buccal cingulum at enamel margin of trigonid; metaconid and protoconid separate, level, connected by ridge at posterior ends; basined talonid showing entoconid set level with hypoconid, ridges extending anterad to bases of protoconid and metaconid.

Fig. 23 *Bassariscus* cf. *B. parvus*, RM_2, ROM 7480, occlusal aspect. Anterior to left, buccal to top of page. Scale 1 mm.

DISCUSSION

The Wood Mountain material has not been compared with the holotype of *Bassariscus parvus*, but comparisons with Merriam's (1916, p. 176) drawing and Hall's photograph (1927, p. 124) show close resemblance. Merriam's measurements for P_4 and M_1 agree well with those of the Wood Mountain specimens (Table 5), except that Merriam's M_1 is 0.7 wider (a difference of almost 15%) but of the same length. The Wood Mountain specimen shows a smaller buccal cingulum than in the holotype of *B. parvus*.

M_2 was not present in the specimen described by Merriam and Hall. The Wood Mountain M_2 is more primitive than any previously described for *Bassariscus* in having an elongate talonid.

The Wood Mountain *Bassariscus*, then, corresponds in size with *B. parvus* but may, with better material, turn out to represent a species slightly more primitive than *B. parvus* in features of M_1 and M_2.

Table 5. Measurements of teeth of *Bassariscus* cf. *B. parvus* compared with those of the holotype of *B. parvus*.

Tooth	Dimension	n	Mean	OR	Holotype of *B. parvus*, UCMP 19768
P_4	AP	2		4.0–4.4	4.9
	BL	2		2.1–2.6	2.4
M_1	AP	1		7.1	7.1
	BL	1		3.1	3.8
M_2	AP	4	4.1	3.9–4.5	
	BL	4	2.3	2.2–2.5	

Family Ursidae

***Ursavus* sp.**
Fig. 24A

REFERRED SPECIMEN. LM^1 (SMNH P76).

DESCRIPTION

Paracone and metacone of similar size, paracone higher, set about equidistant from buccal border; prominent hypocone sending lingual ridge to anterior end of protocone, posterior ridge to posterior end of metaconule; protocone a low ridge, separated posteriorly from metaconule by shallow valley, confluent anteriorly with cingulum on anterior enamel margin of paracone, and with anterior ridge of hypocone; metaconule a low ridge, joining anteriorly transverse ridge from base of metacone, and posteriorly posterior ridge from hypocone; cingulum running around buccal enamel margins of paracone, metacone, anterior enamel margin of paracone.

MEASUREMENTS

M^1, AP(1), 13.7; BL(1), 12.9

DISCUSSION

The tooth is that of an *Ursavus* apparently distinct from *U. pawniensis* in its smaller size. It probably represents an undescribed species close in many respects to *U. primaevus* (Frick, 1926, figs. 42A, 45A).

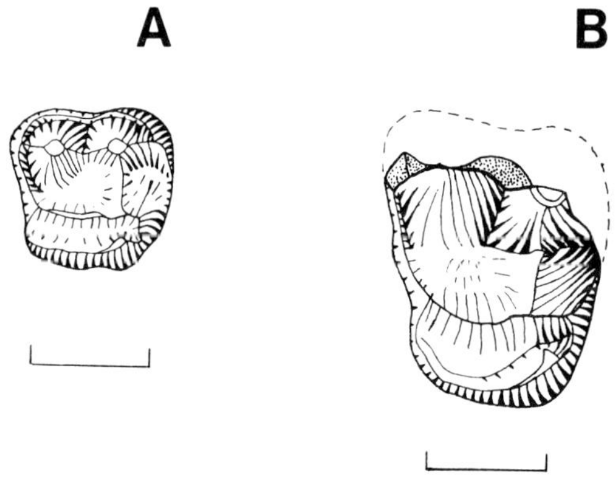

Fig. 24 *Ursavus* sp. and *Hemicyon* sp., upper molars, occlusal aspects. Anterior to left, buccal to top of page. Scales 1 cm.
A. *Ursavus* sp., LM^1, SMNH P76.
B. *Hemicyon* sp., LM^1, SMNH P558.

***Hemicyon* sp.**
Fig. 24B

REFERRED SPECIMEN. LM^1 (SMNH P558).

DESCRIPTION

Tooth broken through paracone and metacone; paracone slightly buccal to metacone; prominent hypocone sending strong lingual cingulum forward to base of protocone; protocone, anteroposteriorly directed ridge, posteriorly joining small metaconule, and anteriorly joining ridge extending anterad forming cingulum at enamel margin of paracone.

DISCUSSION

It does not appear possible to exclude SMNH P558 from the range of variation of *Hemicyon barstowensis* (Frick, 1926, figs. 6A, 7, 8) in either size or pattern.

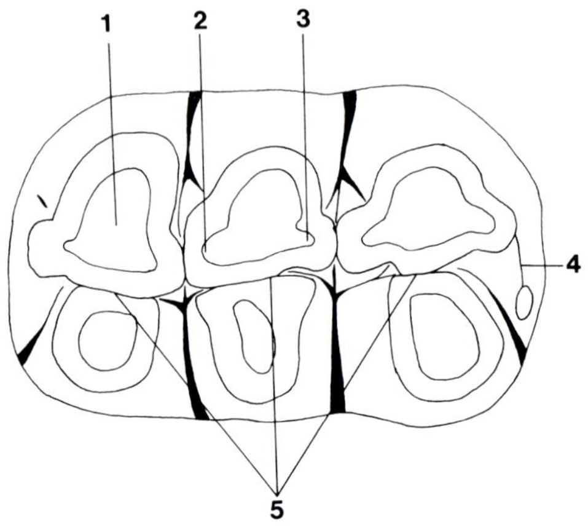

Fig. 25 Gomphotheriidae. RM_1 or M_2 with terminology used in description of lower cheek teeth. Anterior to left, buccal to top of page. Not to scale.
1, hemilophid; 2, anterior ectotrefoil; 3, posterior ectotrefoil; 4, posterior cingulum; 5, median sulcus.

Order Proboscidea
Family Gomphotheriidae

Gomphotheriidae, gen. et sp. indet.
Fig. 26

REFERRED SPECIMENS. RdP_3 (SMNH P378), RM_1 or M_2 (ROM 7606), fragmentary joined left and right mandibles with LM_3, RM_3 (NMC 9149).

DESCRIPTION

Teeth high crowned (DV/BL ratio of last lophid of ROM 7606 = 0.66); enamel of permanent teeth often 5 mm or more thick.

Deciduous premolar (Fig. 26A) relatively small; only posterior lophid preserved; both hemilophids complex, with more wear would appear serrated; posterior spur on buccal hemilophid would form serrated ectotrefoil with wear; strong posterior cingulum bearing numerous cuspules; enamel not thickened as markedly as on permanent molars, maximum thickness of about 3.5 mm shown in SMNH P378.

RM_1 or M_2 (Fig. 26B) made up of three lophids, each composed of two hemilophids; posterior lophid broadest, median and anterior lophids progressively narrower; median sulcus separating hemilophids not deep; simple anterior, posterior ectotrefoils on all lophids.

M_3 (not shown) present only in fragmentary NMC mandibles; only photographs seen, specimen on display and not lent to ROM for study; M_3 tetralophodont, each lophid composed of two hemilophids; median sulcus shallow.

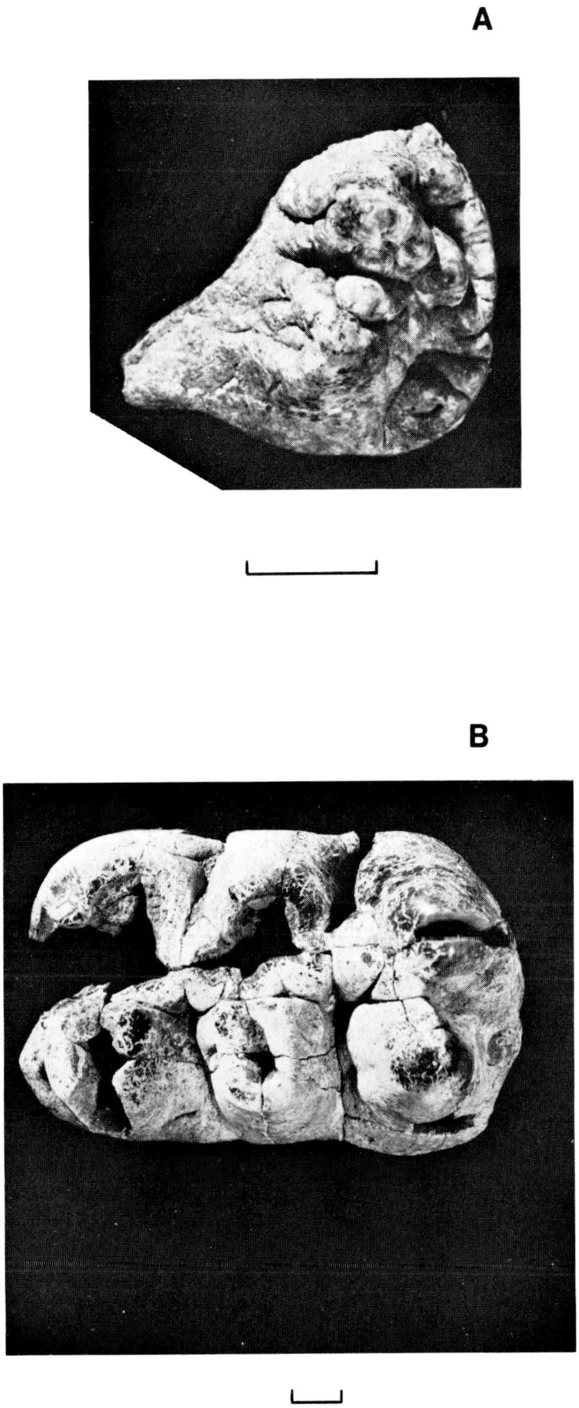

Fig. 26 Gomphotheriidae, gen. et sp. indet., lower cheek teeth in occlusal aspect. Anterior to left, buccal to top of page. Scales 1 cm.

A. RdP_3, SMNH P378.

B. RM_1 or M_2, ROM 7606.

Mandibular symphysis preserved only in most posterior regions; mandible deepening anterior to M_3; ascending ramus arising at approximately centre of fully erupted M_3.

MEASUREMENTS

dP_3, BL(1), 28; DV posterior lophid (1), 22.7. M_1 or M_2, estimated AP(1), 117; BL (1), 72.8; DV posterior lophid (1), 22.7.

DISCUSSION

The specimens described represent a gomphothere similar to many other proboscideans found in Miocene and Pliocene deposits. Characteristics necessary for specific determination were not preserved.

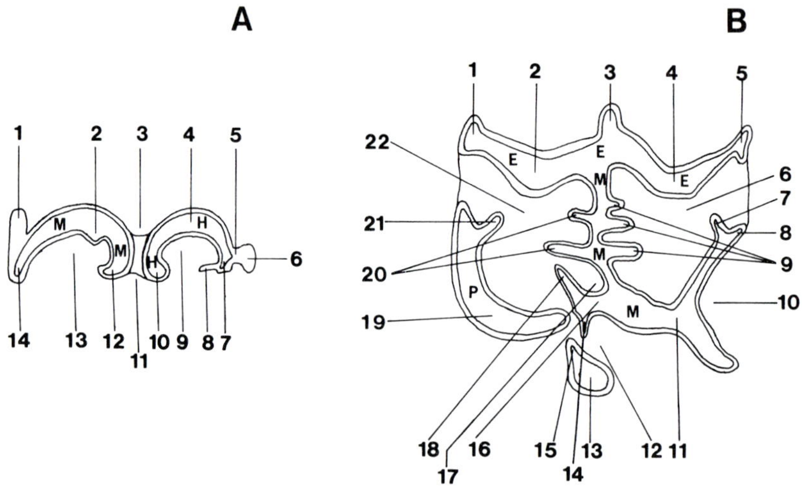

Fig. 27 Equidae. Terminology used in description of cheek teeth. Anterior to left, buccal to top of page. Not to scale.

A. RM_1 or M_2: M, metalophid; H, hypolophid; 1, protostylid; 2, protoconid; 3, ectoflexid; 4, hypoconid; 5, entoconulid (in *Hypohippus*); 6, hypostylid; 7, entoconid; 8, entostylid; 9, postfossettid; 10, mesostylid; 11, entostriid; 12, metaconid; 13, prefossettid; 14, parastylid.

B. LM^1 or M^2: E, ectoloph; P, protoloph; M, metaloph; 1, parastyle; 2, paracone; 3, mesostyle; 4, metacone; 5, metastyle; 6, postfossette; 7, pli hypostyle; 8, hypostyle; 9, pli postfossette; 10, hypofossette; 11, hypocone; 12, medivallum; 13, protocone; 14, pli caballin; 15, spur of protocone; 16, crochet; 17, prefossette loop; 18, pli crochet (or pli protoconule); 19, protoconule; 20, pli prefossette; 21, pli protoloph; 22, prefossette.

Order Perissodactyla
Family Equidae

Merychippus cf. *M. isonesus* (Cope), 1889

Figs. 28, 29, 30

Merychippus praecocidens Russell, 1933

REFERRED SPECIMENS. LP_2 (GSC "Presented by F. Bradshaw, Prov. Mus., Regina"; ROM 1746, 1753), LP_4 (SMNH 7993), LM_1 (SMNH P238, P558; NMC

9045, 9447; GSC XI; ROM 1759, 7545), RM_1 (NMC 8636), LM_2 (SMNH P374, P385, P562), LM_3 (SMNH 7993; GSC XIV), RM_3 (ROM 1744, 7556), joined L. and R. mandibles with LI_{2-3}, RI_{1-3}, LP_{2-4}, RM_1, LM_3 (NMC 9057), L. mandibular fragment with LP_{2-4}, LM_{1-3} (NMC 9064), RP^2 (ROM 7566), LP^4 (NMC 9078; ROM 7609), RM^1 (SMNH P562; NMC 9078—2 specimens; ROM 7586, 7610), LM^2 (NMC 9041; ROM 7590), RM^2 (NMC 9058; ROM 7611), LM^3 (SMNH P566; NMC 8625—paratype of *M. praecocidens*; ROM 7596), RM^3 (SMNH P122, P564; NMC 8624—holotype of *M. praecocidens*), partial skull with LP^{1-4}, RP^1, RP^{3-4}, LM^{1-3}, RM^{1-3} (ROM 7612).

DESCRIPTION

Lower incisors (Fig. 28) arranged in semicircle, procumbent, though not strongly so; roughly oval in cross section; second lower incisor largest of series, third smallest.

P_3 and P_4 broader than lower molars (Fig. 28, list of measurements).

Lower cheek teeth hypsodont, but less so than in *Hipparion* sp. (*vide infra*); greatest height of crown of M_1 (ROM 7545) 29.8, greatest height of crown of M_3 (ROM 7556) 26.0; thin layer of cement, exceeding 1 mm in thickness only within ectoflexid, covering crowns; no cingulum at enamel margin; metaconid and mesostylid separated by pronounced entostriid; protostylid not clearly included in occlusal outline until tooth moderately worn.

Mandibular ramus (Fig. 28) stout; symphyseal region narrowing midlength, expanding anteriorly to incisors; mental foramina midheight on rami, even with posterior end of symphysis, near centres of diastemata; ascending ramus arising posterior to M_3, as continuation of alveolar dorsal border of ramus.

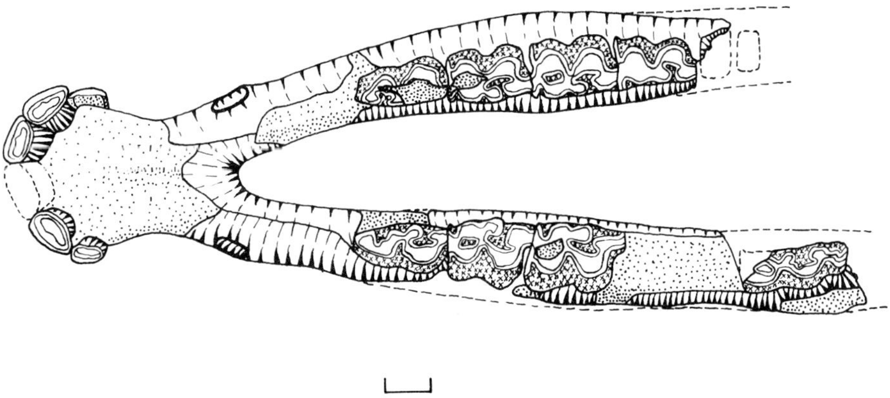

Fig. 28 *Merychippus* cf. *M. isonesus*, joined mandibles with LI_{2-3}, RI_{1-3}, LP_{2-4}, RP_{2-4}, RM_1, LM_3, ROM 9057. Occlusal aspect. Anterior to left, right to top of page. Scale 1 cm.

Upper cheek tooth row (Figs. 29, 30) gently curved, concave lingually; P^1 in normal merychippine shape of lingually flattened oval; P^2 showing anterior expansion of parastyle as in most equids, teeth well worn, pli protoloph and pli hypostyle no longer visible, pli prefossette and pli postfossette weak; metastyles, except on P^4 and M^2, very weak, pli caballins not connected in this late stage of wear to protocones; spur of protocone progressively weaker from anterior to posterior end of tooth row, M^2 and M^3 protocones without spurs; M^3 protocone narrow, with no spur, persistently separated from protoconule, as described for *Merychippus praecocidens* (Russell, 1933, p. 11); wear from mastication producing progressively deeper transverse ridges from anterior to posterior end of tooth row, P^1 and P^2 worn to nearly flat surface and M^2 and M^3 showing deeply ridged occlusal surface.

Upper cheek teeth, like lowers, hypsodont but less so than in *Hipparion* sp.; greatest height of crown of M^3 (SMNH P564) 36.1; a thin layer of cement present, no cingula at enamel margins.

Hypofossettes of upper cheek teeth deep, extending to within 8 mm of roots; pli caballins short, sometimes bifid, extending into medivalla but not reaching protocones except in extremely worn teeth; pli pre- and postfossettes bifid or trifid; marked pli hypostyle may be shown, but no prominent pli protoloph, perhaps because few specimens assigned to *M.* cf. *M. isonesus* heavily worn; prefossette loops not extending linguad as far as lingual borders of pre- and postfossettes; parastyles, mesostyles strong at all stages of wear, metastyles usually becoming prominent only in medium and late wear stages.

Skull, ROM 7612 (Figs. 29, 30), somewhat crushed, left side displaced anterad about 15 mm; individual probably mature, sutures, fully closed, not distinguishable from cracks caused by distortion; skull apparently comparatively high, narrow, but this appearance probably created by lateral crushing; palate narrowed considerably by crushing (Fig. 29).

Malar ridge (Fig. 29) broad; posterior palatine foramen even with posterior portion of M^2 on each side, palatal vacuities terminating anteriorly at about centre of M^2; incisive foramina not shown, not extending posterad as far as anterior end of P^1.

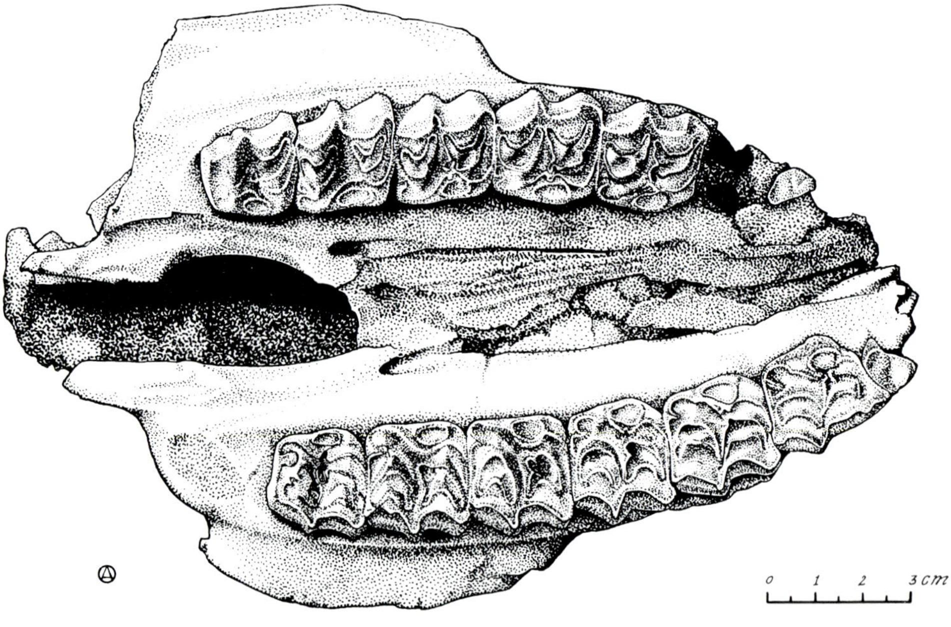

Fig. 29 *Merychippus* cf. *M. isonesus*, partial skull, ROM 7612, palatal aspect. Anterior to right, right to top of page. Scale in cm.

Temporal table (Fig. 30) crushed; malar ridge extending anterad to posterior end of P^4; malar facial fossa dorsally deep and well defined, not extending far anterad, ending about 30 mm dorsal to malar ridge; the malar fossa extending posterad to centre of M^2, connected to deep lachrymal fossa terminating anteriorly at centre of P^3; deep infraorbital foramen, opening anterad, situated about 10 mm ventral to lachrymal fossa, directly dorsal to anterior end of P^4.

MEASUREMENTS

I_1, AP(1), 9.8; BL(1), 6.2. I_2, AP(2), OR 11.0–11.2; BL(2), OR 5.8–6.2. I_3, AP(2), OR 9.3–9.5; BL(1), 5.1. P_2, AP(5), mean 20.5 ± .3, OR 19.1–21.3, CV 3.4; BL(5), mean 11.2 ± .3, OR 10.3–12.2, CV 6.3. P_3, AP(3), mean 19.2, OR 19.2–19.3; BL(3), mean 11.7, OR 11.4–12.0. P_4, AP(3), mean 19.3, OR 19.1–19.4; BL(4), mean 12.0, OR 11.2–13.6. M_1, AP(10), mean 20.0 ± .3, OR 18.7–21.9, CV 5.0; BL(10), mean 10.4 ± .2, OR 9.6–11.6, CV 6.7. M_2, AP(4), mean 19.4, OR 17.1–23.1; BL(4), mean 10.7, OR 9.1–11.4. M_3, AP(3), mean 22.7, OR 21.8–24.0; BL(5), mean 8.8 ± .5, OR 7.9–10.7, CV 11.4. P^1, AP(2), OR 10.5–10.7; BL(2), OR 6.3–7.3. P^2, AP(2), OR 21.1–25.9; BL(2), OR 18.6–20.4. P^3, AP(2), OR 22.2–22.3; BL(2), OR 23.3–23.4. P^4, AP(4), mean 22.0, OR 20.7–22.8; BL(4), mean 23.0, OR 22.7–23.5. M^1, AP(6), mean 20.1 ± .1, OR 19.8–20.6, CV 1.5; BL(7), mean 21.4 ± .4, OR 19.4–22.7, CV 5.6. M^2, AP(6), mean 20.2 ± .4, OR 18.6–21.6, CV 4.5; BL(5), mean 21.9 ± .4, OR 20.5–23.1, CV 4.1. M^3, AP(6), mean 18.8 ± .8, OR 17.1–21.7, CV 10.6; BL(8), mean 18.5 ± .4, OR 17.3–20.8, CV 7.0.

DISCUSSION

Isolated teeth were assigned to relative positions using the method of Bode (1931).

Extensive comparisons of Wood Mountain material with specimens in the Frick Laboratory, American Museum of Natural History, showed that two taxa

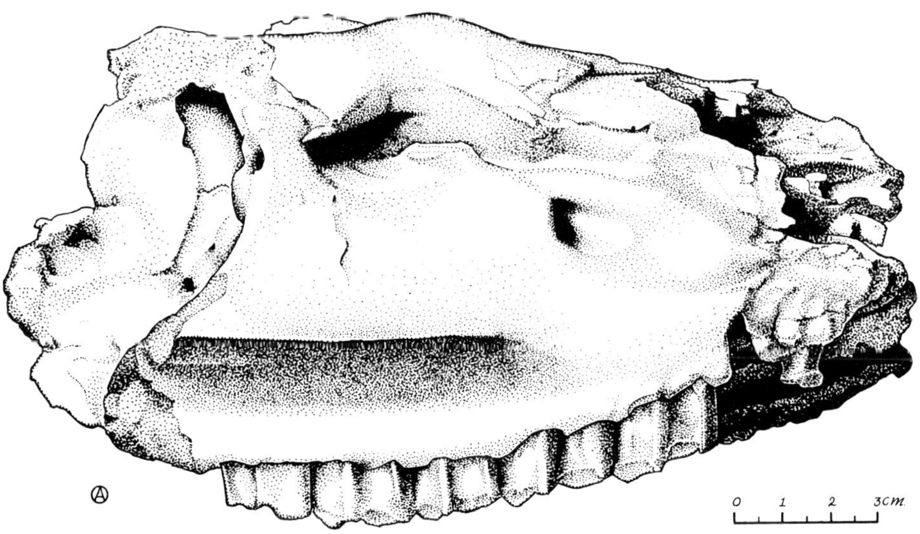

Fig. 30 *Merychippus* cf. *M. isonesus*, partial skull, ROM 7612, lateral aspect. Anterior to right, dorsal to top of page. Scale in cm.

of hypsodont equids are represented in the Wood Mountain Fauna, *Merychippus* and *Hipparion*. Only unworn or little-worn isolated teeth can be definitely assigned to one genus or the other, as the most important difference between the taxa is the difference in the degree of hypsodonty. Thus most of the isolated teeth are indeterminate between *Merychippus* and *Hipparion*. There is no criterion of size of the occlusal surface by which isolated teeth of *Merychippus* can be distinguished from corresponding teeth of *Hipparion*, and the entire sample of teeth of hypsodont equids demonstrates a size range that might be seen in a natural population.

Lower cheek teeth of *Merychippus* from Wood Mountain are distinguished from corresponding teeth of *Hipparion* by their shorter height of crown, consistent absence of a pli caballinid, and late-developing protostylid. Upper cheek teeth can be dependably separated from corresponding teeth of *Hipparion* only by their lesser degree of hypsodonty.

Skinner (pers. comm., 1969) has stated that the partial *Merychippus* skull (ROM 7612) belongs within *Merychippus* cf. *M. isonesus*. The shape, depth, and size of the facial fossae in the Wood Mountain skull are well within the range of variation of specimens assigned to *Merychippus* cf. *M. isonesus* from Nebraska, and the characteristics of the teeth correspond well to specimens of *M.* cf. *M. isonesus*. Measurements of teeth of AMNH 8175, holotype of *Merychippus isonesus* (estimated from Osborn, 1918, fig. 75) are: P^1, AP, 12; BL, 6.5. P^2, AP, 22.5; BL, 17. P^3, AP, 20.5; BL, 20.5. P^4, AP, 20; BL, 19.5. M^1, AP, 19; BL, 19. M^2, AP, 21; BL, 18.5. M^3, AP, 17.5; BL, 13.5. The Wood Mountain material can be assigned to *M.* cf. *M. isonesus*, which is represented by many skulls from the Great Plains and Great Basin, and thus is the most common and best-known equid taxon of the Great Plains Upper Miocene.

Merychippus praecocidens Russell, 1933, is regarded here as a junior synonym of *M. isonesus*. The holotypic and paratypic M^3s correspond well in important characters to the M^3s of ROM 7612, the skull referred to *M.* cf. *M. isonesus* on characters of the teeth and facial fossae.

Hipparion sp.

REFERRED SPECIMENS. LP_4 (SMNH P564, P566; NMC 9045; ROM 7535, 7537, 7540), RP_4 (NMC 9045, 9411; ROM 7533, 7534), LM_1 (ROM 7527), RM_1 (SMNH P76, P566; NMC 9045, 9390; ROM 7541, 7551), RM_2 (SMNH P551), LM_3 (SMNH P377, 7993), RM_3 (NMC 9411; ROM 7555, 7557), RP^3 (SMNH P123; ROM 7572), RP^4 (ROM 7576), RM^1 (NMC 9041), LM^2 (SMNH 7988, 7995), LM^3 (ROM 7598), RM^3 (SMNH 7990).

DESCRIPTION

Cheek teeth hypsodont, bearing no cingula at enamel margins, covered by thin layer of cement never more than 1 mm thick except in ectoflexid or medivallum.

Lower cheek teeth far more hypsodont than corresponding teeth of *Merychippus*; greatest height of crown of M_2 (SMNH P551) 49.3, greatest height of crown of M_3 (SMNH P377) 36.7; metaconids, mesostylids separated by pronounced entostriids; protostylid included in occlusal surface in early wear; pli caballinid well developed in three specimens of P_4 (NMC 9045, 9411; ROM 7540); hypostylid expanding markedly toward base of crown.

Greatest height of crown of M^2 (SMNH 7995) 44.7; hypofossette extending to within 5 mm of roots in upper cheek teeth.

Pli caballin stronger than in *Merychippus* cf. *M. isonesus*, in three moderately worn specimens nearly reaching protocone; prefossette loop extending linguad past lingual borders of pre- and postfossettes in one specimen (ROM 7576), but not in another (NMC 9041); pli prefossette bifid or trifid, and in ROM 7576, pli postfossette with four branches; individual branches of pli pre- and postfossettes may themselves bifurcate; as in *M.* cf. *M. isonesus*, metastyle usually not prominent until tooth relatively worn.

MEASUREMENTS

P_4, AP(9), mean 21.2 ± .3, OR 19.2–22.7, CV 4.7; BL(10), mean 12.0 ± .3, OR 10.5–12.8, CV 8.3. M_1, AP(6), mean 21.2 ± .5, OR 19.7–22.8, CV 5.7; BL(7), mean 10.5 ± .3, OR 9.6–11.8, CV 7.6. M_2, AP(1), 22.2; BL(1), 11.1. M_3, AP(4), mean 23.6, OR 21.0–25.0; BL(5), mean 9.5 ± .1, OR 8.9–9.7, CV 3.2. P^3, AP(2), OR 21.2–21.7; BL(2), OR 19.6–20.2. P^4, AP(1), 21.8; BL(1), 22.1. M^1, AP(1), 19.6; BL(1), 22.4. M^2, AP(2), OR 20.9–23.2; BL(2), OR 21.0–21.2. M^3, AP(2), OR 19.1–20.1; BL(1), 19.8.

DISCUSSION

The teeth listed above were assigned relative positions by the method of Bode (1931).

Lower teeth of *Hipparion* sp. from the Wood Mountain Formation are distinguished from corresponding teeth of *Merychippus* by their greater degree of hypsodonty, earlier development of the protostylid, and occasional presence of a pli caballinid. Upper cheek teeth are marked by their greater degree of hypsodonty.

The degree of hypsodonty of the specimens referred to *Hipparion* sp. from the Wood Mountain Formation corresponds to the degree of hypsodonty seen in primitive species of *Hipparion* in the Frick Collection, American Museum of Natural History. But these teeth cannot be assigned to any particular species.

The presence of teeth of a primitive *Hipparion* at widely scattered localities in the Wood Mountain Formation gives added strength to the temporal assignment made on the bases of the rodent, carnivore, and artiodactyl assemblages. Rodents of the Wood Mountain Fauna indicate an age close to that of the Norden Bridge local fauna, basal Valentine Formation of Nebraska, whereas the similarity of the carnivore and ungulate taxa to Lower Snake Creek forms indicates that the age is greater than that of the Norden Bridge Fauna. Skinner (pers. comm., 1969) stated that the earliest occurrence of *Hipparion* in Nebraska is at the base of the Valentine Formation. It is certainly plausible that *Hipparion* should be found in a fauna just older than the basal Valentine Formation. The presence of both *Merychippus* and *Hipparion* in the Wood Mountain Fauna, far from being an indication of faunal mixing, corroborates the age determination made with other criteria.

Merychippus or *Hipparion*

REFERRED SPECIMENS. LdP_2 (NMC 9081), LP_2 (SMNH P557, P568; NMC 9411; ROM 7529), RP_2 (SMNH 7993, 7995; NMC 9390; GSC V; ROM 1737, 1756, 7516, 7517, 7521), LdP_3 (NMC 9046), RdP_3 (ROM 7530), LP_3 (SMNH P560, 7993, ROM 7520), RP_3 (SMNH P566; ROM 7518, 7519, 7526, 7529), LP_4 (SMNH P122, P236, P237, P378, P562—2 specimens; NMC 9045—2 specimens; ROM 7531, 7532, 7538, 7547), $RP_{\bar{4}}$ (SMNH P123, P378, P550, P562, P568, 8494; NMC 9079, 9411—2 specimens; ROM 7536, 7539), RP_3 or P_4 (ROM 7522), R. mandibular fragment with RdP_{2-4} (NMC 8983), L. mandibular fragment with LP_{3-4}

(ROM 7524, 7525), LM$_1$ (SMNH P562, 7993, 8494; NMC 9045—2 specimens, 9059), RM$_1$ (SMNH P236, P550, P564, 7993; NMC 9045—2 specimens; ROM 7544, 7546, 7550, 7562), L. mandibular fragment with LP$_{2-4}$, LM$_1$ (ROM 7523), L. mandibular fragment with broken LP$_{3-4}$, LM$_1$ (SMNH 7998), LM$_2$ (SMNH P380, P562—2 specimens, 8494), RM$_2$ (SMNH P122, P557—2 specimens; NMC 8635; ROM 7542, 7543, 7552, 7553), LM$_3$ (SMNH P123—2 specimens, P373, P557; NMC 9411; ROM 7548, 7549, 7554), RM$_3$ (SMNH P122, P566, 7991, 7992; NMC 9045; ROM 1760, 7558–7561), LP^2 (SMNH P123—2 specimens, P378, P562, 5278; NMC 9041, 9390; ROM 7563, 7564), RP^2 (SMNH P384, P551), LdP^3 (NMC 9042), RdP^3 (NMC 9411—2 specimens), LP^3 (NMC 9041, 9390—3 specimens, 9411; ROM 1735, 7569), RP^3 (SMNH 7993, 7995; NMC 9041, 9390; ROM 7570, 7573, 7574, 7599), RdP^4 (NMC 9042), LP^4 (SMNH P551, P562, P566, 7993; NMC 9390; ROM 7575, 7577, 7578), RP^4 (SMNH P123, P765; ROM 7579–7584), LM^1 (SMNH P391—2 specimens, P564; NMC 8637, 9041—2 specimens, 9078; ROM 7585), RM^1 (SMNH P237; ROM 7587–7589), LM^2 (SMNH P550, P566; NMC 9411; ROM 7591, 7592), RM^2 (SMNH P76, P379; NMC 9411; ROM 7593–7595), LM^3 (ROM 1758, 7597).

DESCRIPTION

Lower deciduous premolars (not shown) far lower crowned than permanent premolars of either *Merychippus* cf. *M. isonesus* or *Hipparion* sp. from Wood Mountain Formation; hypostylids high, prominent; metaconids, mesostylids separated by wide, deep entostriids; buccal enamel margins blocked by basal stylids; no cement on dP$_2$, thin coating of cement filling buccal opening of ectoflexid of dP$_3$, cement covering much of buccal surface of dP$_4$. Upper deciduous premolars (not shown) differing from permanent premolars of *Merychippus* and *Hipparion* in extreme brachydonty.

Other lower and upper cheek teeth indeterminate between *Merychippus* and *Hipparion* as described above, heavy wear removing diagnostic features.

MEASUREMENTS

dP$_2$, AP(2), OR 22.5–23.2; BL(2), OR 10.4–12.0. P$_2$, AP(12), mean 21.1 ± .2, OR 20.0–22.8, CV 3.8; BL(13), mean 11.7 ± .3, OR 10.1–13.9, CV 7.7. dP$_3$, AP(3), mean 20.5, OR 17.6–22.4; BL(3), mean 12.4, OR 11.8–12.9. P$_3$, AP(7), mean 19.3 ± .5, OR 17.6–22.4, CV 7.3; BL(8), mean 12.7 ± .4, OR 11.1–14.7, CV 7.9. dP$_4$, BL(1), 12.8. P$_4$, AP(23), mean 19.2 ± .3, OR 16.7–21.6, CV 6.8; BL(23), mean 12.9 ± .2, OR 11.5–14.6, CV 6.2. M$_1$, AP(16), mean 19.7 ± .5, OR 16.4–22.6, CV 9.1; BL(15), mean 11.3 ± .2, OR 9.8–12.9, CV 8.8. M$_2$, AP(12), mean 18.9 ± .3, OR 17.4–20.4, CV 4.8; BL(9), mean 11.0 ± .4, OR 9.3–12.7, CV 10.0. M$_3$, AP(16), mean 22.3 ± .3, OR 20.2–24.2, CV 5.4; BL(17), mean 9.9 ± .2, OR 8.8–11.3, CV 8.1. P^2, AP(11), mean 24.5 ± .4, OR 22.5–26.9, CV 6.1; BL(11), mean 19.2 ± .4, OR 17.3–21.9, CV 7.8. dP^3, AP(3), mean 21.7, OR 19.2–24.2; BL(2), OR 18.4–19.2. P^3, AP(14), mean 20.1 ± .4, OR 18.3–22.0, CV 7.0; BL(13), mean 21.3 ± .4, OR 18.4–23.7, CV 7.5. dP^4, AP(1), 21.1; BL(1), 20.3. P^4, AP(15), mean 16.9 ± .4, OR 16.9–22.3, CV 9.5; BL(14), mean 21.3 ± .4, OR 18.4–23.7, CV 6.6. M^1, AP(9), mean 22.2 ± .5, OR 18.0–23.0, CV 7.2; BL(12), mean 22.3 ± .3, OR 20.7–24.9, CV 4.9. M^2, AP(11), mean 20.0 ± .3, OR 18.6–21.5, CV 5.0; BL(9), mean 21.0 ± .8, OR 19.2–28.0, CV 11.9. M^3, AP(2), OR 16.7–18.7; BL(2), OR 17.0–19.1.

DISCUSSION

Lower deciduous premolars differ from those referred to *Hypohippus* cf. *H. osborni* (*vide infra*) chiefly in their smaller size, more prominent entostriids, presence of a basal stylid in the ectoflexid, and presence of cement on dP_3 and dP_4. Upper deciduous premolars are distinguished from those of *Hypohippus* by their strong crochets and the presence of minute quantities of cement in the fossettes and medivalla.

Cheek teeth not in mandibular fragments have been assigned relative positions using Bode's criteria (1931).

Archaeohippus cf. *A. minimus* (Douglass), 1899

Fig. 31

REFERRED SPECIMENS. RM_1 or M_2 (GSC II), L. upper cheek tooth (ROM 7692), R. upper cheek tooth (GSC III, IV).

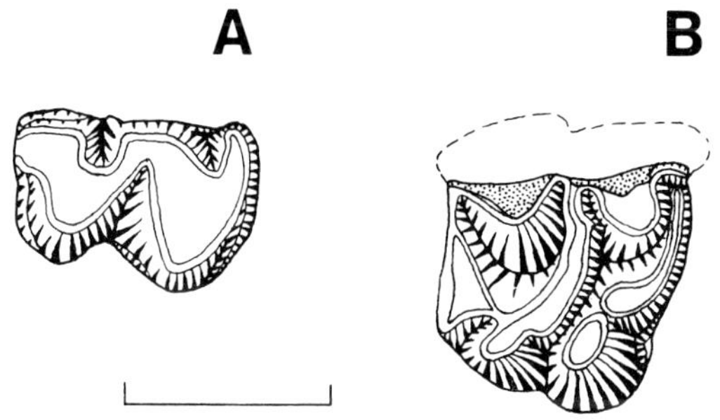

Fig. 31 *Archaeohippus* cf. *A. minimus*, cheek teeth. Occlusal aspects. Anterior to right. Scale 1 cm.
A. RM_1 or M_2, GSC II, buccal to bottom of page.
B. R. upper cheek tooth, GSC III, buccal to top of page.

DESCRIPTION

Small equid, cheek teeth brachydont, cement-free.

Lower molar (Fig. 31A) worn, enamel surface pitted; well-defined cingulum at anterior enamel margin, another cingulum at posterobuccal enamel margin; small cingulum at buccal enamel margin of protoconid, ending at ectoflexid; short lingual cingula at enamel margin of entoconid and base of prefossettid; metaconid, mesostylid not separated by entostriid in GSC II, cannot be distinguished.

Upper cheek teeth (Fig. 31B) showing protoloph and metaloph separated by open prefossette and remaining separate during wear; no crochet on metaloph; hypostyle joining metaloph early; protoloph reaching parastyle only in well-worn specimens.

DISCUSSION

Distinctive small size, weak development of cingula, and distinctness of the protoloph from the parastyle indicate that the specimens considered here are referable to *Archaeohippus*. Absence of a crochet indicates that the Wood Mountain form belongs to the group of *Archaeohippus* showing anchitherine resemblances.

Of the described forms lacking crochets, *Archaeohippus* from the Wood Mountain Formation differs from *A. penultimus* in the presence of both buccal and lingual cingula on lower molars. *A. ultimus* shows some resemblances, but the cingula of the lower teeth are stronger than in the Wood Mountain material, and cement is occasionally present (Downs, 1956, p. 247). The closest similarity is to *A. minimus*, which shows comparable development of the cingula of the lower cheek teeth, comparable size (Table 6), and comparable development of the protoloph.

Table 6. Measurements of teeth of *Archaeohippus* cf. *A. minimus* compared with those of holotype (CM 713) and paratype (CM 714) of *A. minimus* (estimated from Osborn, 1918, fig. 72a).

Tooth	Dimension	n	OR	Paratype and holotype
Lower molar	AP	1	12.2	11.5
	BL	1	8.2	9
Upper cheek teeth	AP	2	12.1–14.5	13.5
	BL			17.5

Hypohippus cf. *H. osborni* Gidley, 1907
Figs. 32, 33

REFERRED SPECIMENS. LdP_2 (NMC 9080), L. lower premolar or molar (NMC 9412), R. lower premolar or molar (NMC 9388), R. mandibular fragment with RdP_{2-4}, RM_2 (NMC 9479), LM_3 (NMC 9080), RM_3 (NMC 9412), L. mandibular fragment with LP_4, LM_{1-3} (SMNH P568), RP^2 (ROM 7605), R. upper premolar or molar (ROM 7604).

DESCRIPTION

Deciduous and permanent cheek teeth brachydont; no cement on teeth except for thin layer on part of lingual surface of upper cheek tooth ROM 7604.

In deciduous lower premolars (FIG. 32A), metaconid and mesostylid separated by shallow entostriid not persisting to base of crown; buccal cingula on dP_2 to dP_4; no basal stylid blocking ectoflexid.

In permanent lower molars (Fig. 33A), buccal cingulum strong; M_3 ectoflexid may be blocked by weak basal stylid; prefossettid blocked by short lingual cingulum; entostriid very weak; in M_3, small entoconulid at connection between entoconid and hypostylid.

Horizontal ramus (Fig. 32B) relatively shallow, ventral margin making small angle with tooth row; single mental foramen beneath alveolus for dP_1.

In upper cheek teeth (Fig. 33B), metaloph joining ectoloph, connected in early stages of wear to large, triangular hypostyle; no crochet on P^2, metaloph and protoloph separated by open prefossette; protoloph separated by deep valley from parastyle, joining first to paracone; lingual cingula on lingual two-thirds of anterior and posterior faces and across open prefossette; short buccal cingula anterobuccal to paracone and posterolingual to metacone.

DISCUSSION

NMC 9679 shows the same size of teeth, patterns of deciduous lower premolars, and shape, depth, and length of jaw as AMNH 8256, referred to *Hypohippus osborni* from the Pawnee Creek Beds of Colorado. The anterior portion of M_2 (Fig. 32B) shows great similarity to M_2 of AMNH 8256.

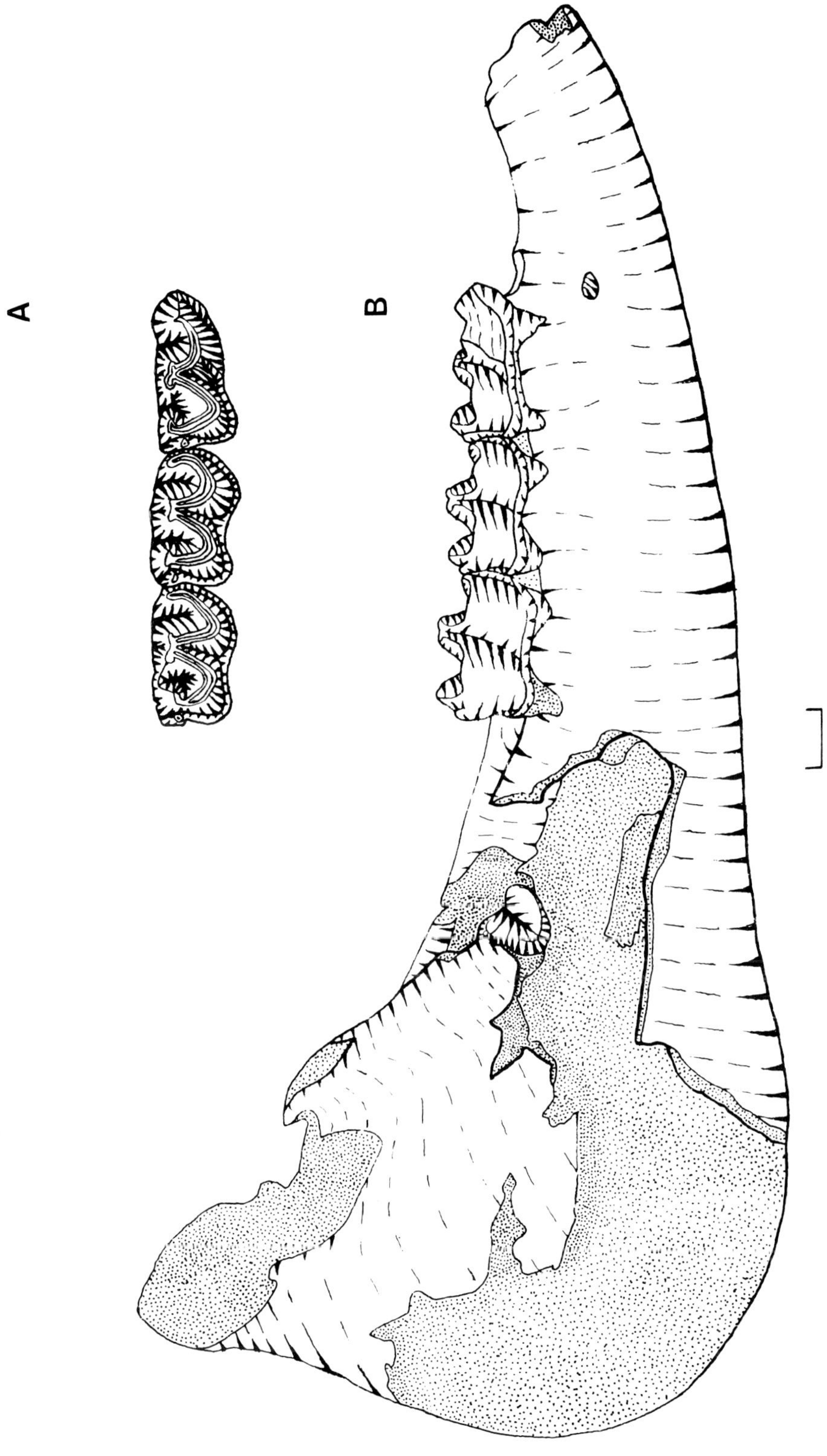

Fig. 32 *Hypohippus* cf. *H. osborni*, R. mandible with RdP$_{2-4}$, RM$_2$, NMC 9479. Anterior to top of page. Scale 1 cm.

A. Occlusal aspect, buccal to right of page.

B. Lateral aspect, occlusal to left of page.

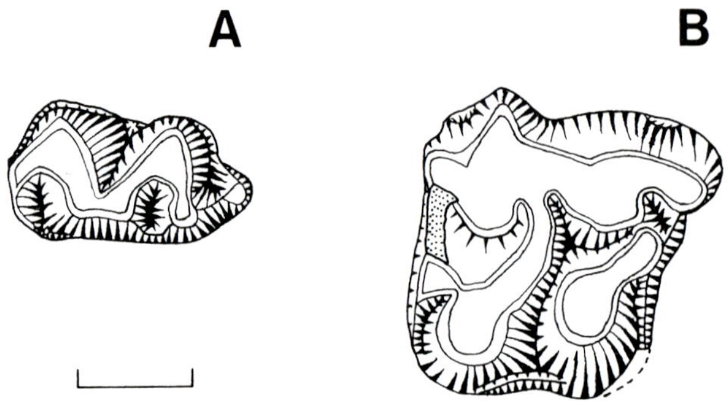

Fig. 33 *Hypohippus* cf. *H. osborni*, cheek teeth, occlusal aspects. Buccal to top of page. Scale 1 cm.
A. RM_3, NMC 9412, anterior to left.
B. RP^2, ROM 7605, anterior to right.

The holotype of *H. osborni*, on display in the American Museum of Natural History, was not available for detailed comparison; Wood Mountain teeth agree in size with the holotype (Table 7). AMNH 9395, the paratype, from the "Loup Fork" of Colorado, matches NMC 9479 in the size of M_2. SMNH P568 compares closely with this specimen in the length of the molar series and characteristics of the molars.

ROM 7605, RP^2, corresponds to AMNH 9395 in all features, including loph connections and the strength of the lingual cingulum.

Table 7. Measurements of teeth of *Hypohippus* cf. *H. osborni* compared with those of the holotype of *H. osborni* (estimated from Osborn, 1918, fig. 168).

Tooth	Dimension	n	OR	Holotype of *H. osborni*, AMNH 9407
dP_2	AP	2	26.7–27.0	
	BL	2	14.0–14.5	
dP_3	AP	1	23.5	
	BL	1	14.1	
dP_4	AP	1	24.7	
	BL	1	14.6	
M_2	AP	1	26.9	25
	BL	1	20.9	
Lower molar or premolar	AP	1	26.4	
	BL	1	18.4	
M_3	AP	2	28.7–29.7	27
	BL	2	16.1–17.1	
P^2	AP	1	26.0	28
	BL	1	25.7	25.5
Upper molar or premolar	AP	1	21.6	
	BL	1	20.8	

?Anchitherinae, gen. et. sp. indet.

Fig. 34A

REFERRED SPECIMEN. L. lower molar (SMNH P568).

DESCRIPTION

Tooth (Fig. 34A) in advanced stage of wear (crown height 10.3); no cingula at enamel margins; cement may have covered crown, traces preserved in ectoflexid and postfossettid, and anterobuccal to protoconid and posterobuccal to hypoconid.

Metaconid and mesostylid separated by deep entostriid continuing to base of crown; postfossettid open and prominent to within 3.5 mm of base of crown.

Protostylid prominent; hypostylid expanding at base of crown to traverse entire width of crown.

DISCUSSION

It is not clear to which taxon this tooth should be referred. There is a superficial similarity to *Merychippus* cf. *M. isonesus*, but *M.* cf. *M. isonesus* does not show the prominent open entostriid and postfossettid, or basal lateral expansion of the hypostylid. In *Hipparion* sp., the prominent open entostriid and postfossettid are not shown. Moreover, SMNH P568 appears to represent a far lower-crowned tooth than any example of *Hipparion* sp. from the Wood Mountain Fauna.

Hypohippus and *Archaeohippus* lack the prominent entostriid, and in both taxa from the Wood Mountain Fauna the prefossettid is open to the same level as the postfossettid. The Wood Mountain *Archaeohippus* has basal cingula and lacks cement, and neither taxon shows basal expansion of the hypostylid to the same degree as in SMNH P568.

Only *Hippodon speciosus* agrees with SMNH P568 in all characters including size. Comparative measurements of SMNH P568 and AMNH 465, the holotype of *Hippodon speciosus*, are given in Table 8. SMNH P568 shares with *Hippodon* the closing of the postfossettid after the prefossettid, the deep and persistent entostriid, and the basal lateral expansion of the hypostylid. The specimen of *Hippodon* lacks cement, perhaps because it was in an early stage of development.

It has generally been agreed that it is best if specimens are not referred to *Hippodon*, leaving it "a monotypic genus, *nomen dubium*, with a single species

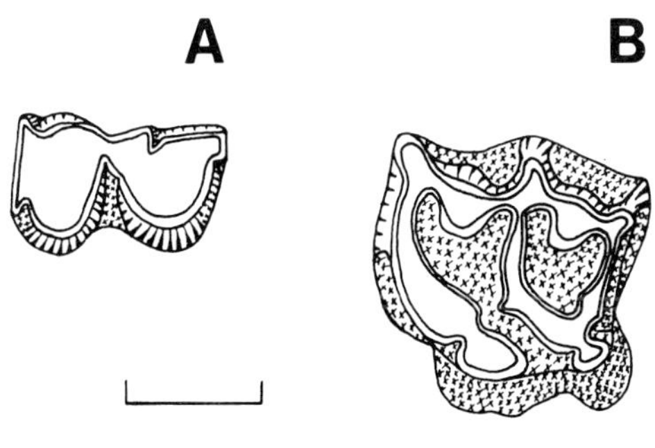

Fig. 34 ?Anchitherinae, gen. et sp. indet. and Equidae, gen. et sp. indet., cheek teeth in occlusal aspect. Scale 1 cm.

A. ?Anchitherinae, gen. et sp. indet., L. lower molar, SMNH P568, anterior to right, buccal to bottom of page.

B. Equidae, gen. et sp. indet., LM^1, NMC 9411, anterior to left, buccal to top of page.

Table 8. Measurements of unidentified ?anchitherine tooth (SMNH P568) compared with those of cast (ROM 7699) of holotype of *Hippodon speciosus* (AMNH 465).

Dimension	SMNH P568	Holotype of *Hippodon speciosus*
AP (maximum)	15.8	18.6
AP 10 mm above base of crown	15.8	17.5
BL	10.3	10.9

standing for its type" (Skinner and Taylor, 1967, p. 17). Thus the Wood Mountain specimen is not given a precise designation, although it closely resembles only *Hippodon speciosus*.

Equidae, gen. et sp. indet.

Fig. 34B

REFERRED SPECIMEN. LM^1 (NMC 9411).

DESCRIPTION

Tooth well worn but hypsodont; crochet weakly developed, protoloph and metaloph separated by deep prefossette; no pli protoloph, pli prefossette or pli postfossette; pli hypostyle weak; parastyle and mesostyle strong; metastyle well developed; protocone and protoconule strongly connected; hypostyle connected to metastyle; hypofossette deep, reaching to within 9 mm of base of crown; thin layer of cement covering crown.

MEASUREMENTS

M^1, AP(1), 19.8; BL(1), 19.7.

DISCUSSION

NMC 9411 is apparently not similar to any described species. The only reported specimen to which comparisons can be made is the badly broken tooth referred to *Protohippus parvulus* by Marsh (Osborn, 1918, fig. 105), which shares with NMC 9411 the overall lack of plis, strong connection of the protocone and protoconule, and weak development of the crochet.

Family Rhinocerotidae

?*Aphelops* sp.

Fig. 36

REFERRED SPECIMENS. LP_2 (SMNH P123; ROM 7602), RP_2 (ROM 7600), LP_3 or P_4 or M_1 or M_2 (SMNH P123, P1159; NMC 9407), RP_3 or P_4 or M_1 or M_2 (NMC 8973), R. mandibular fragment with RP_3 or P_4 or M_1 or M_2 (NMC 9480), RM^1 (ROM 7601), RM^2 (ROM 1734), upper molar (SMNH P557).

DESCRIPTION

P_2 (Fig. 36A) proportionally shorter, broader than other, massive, lower cheek teeth (Fig. 36B); P_3 to M_3 with cingula anterobuccal to paraconid, posterior to hypoconid and entoconid; P_2 showing cingulum at buccal enamel margin, buccal to hypoconid and protoconid; lower cheek teeth mesodont.

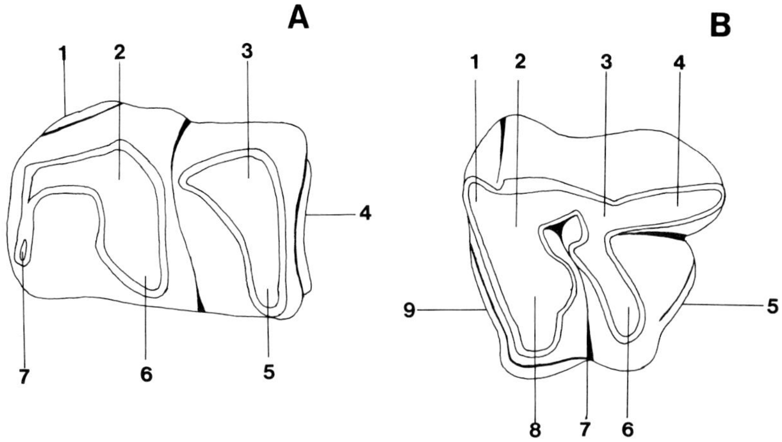

Fig. 35 ?*Aphelops* sp. Terminology used in description of cheek teeth. Anterior to left, buccal to top of page. Not to scale.

A. RM_1 or M_2: 1, anterobuccal cingulum; 2, protoconid; 3, hypoconid; 4, posterior cingulum; 5, entoconid; 6, metaconid; 7, paraconid.

B. LM^2: 1, parastyle; 2, paracone; 3, metacone; 4, metastyle; 5, posterior cingulum; 6, hypocone; 7, entoflexus; 8, protocone; 9, anterolingual cingulum.

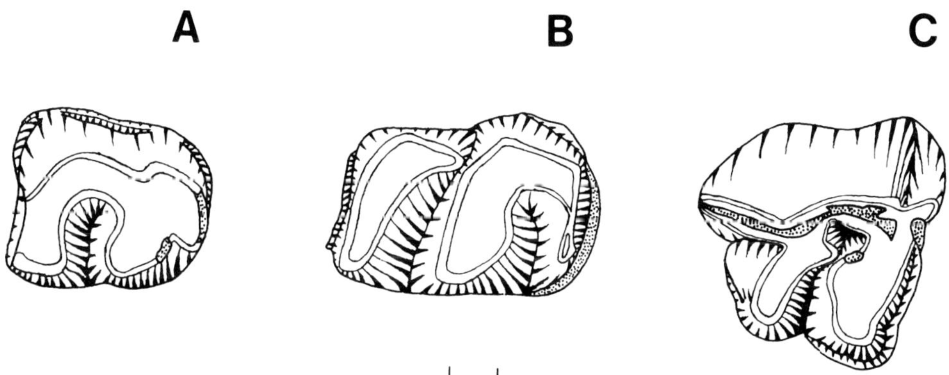

Fig. 36 ?*Aphelops* sp., cheek teeth in occlusal aspect. Anterior to right, buccal to top of page. Scale 1 cm.

A. LP_2, ROM 7602.

B. LP_3 or P_4 or M_1 or M_2, NMC 9407.

C. RM^2, ROM 1734.

M^2 (Fig. 36C) with strongly developed crochet; strong anterolingual, posterior cingula; mesodont.

No cement present on teeth.

MEASUREMENTS

P_2, AP(1), 41.3; BL(1), 36.1; DV(1), 26.4. P_3 or P_4 or M_1 or M_2, AP(3), mean 49.0, OR 41.2–55.5; BL(3), mean 31.6, OR 28.6–37.0. M^2, AP(1), 49.6; BL(1), 49.8; DV(1), 34.0.

DISCUSSION

Relatively complete ramal or cranial specimens of rhinocerotids are necessary for reliable identification. *Aphelops* and *Teleoceras* possess similar teeth, but *Aphelops* is usually considered to have relatively short-crowned cheek teeth (Osborn, 1904, p. 311). These Wood Mountain teeth are of about the size and height of crown of *Aphelops ceratorhinus* (Osborn, 1904, pp. 316, 317).

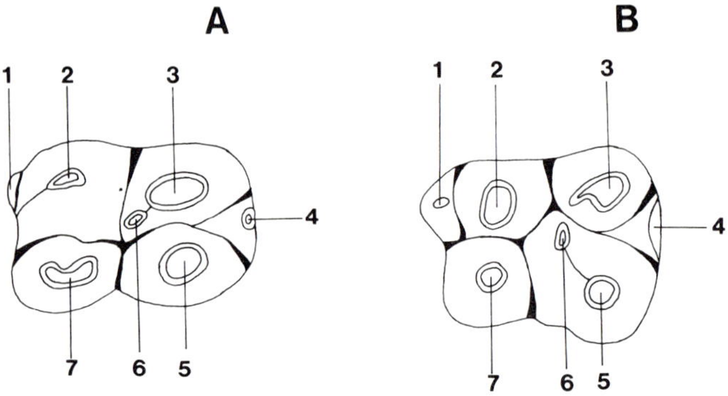

Fig. 37 Tayassuidae. Terminology used in description of cheek teeth. Anterior to left, buccal to top of page. Not to scale.

A. RM_1 or M_2: 1, anterior accessory cuspule; 2, protoconid; 3, hypoconid; 4, posterior accessory cuspule; 5, metaconid; 6, mesoconid; 7, paraconid.

B. LM^1 or M^2: 1, anterior accessory cuspule; 2, paracone; 3, metacone; 4, posterior cingulum; 5, hypocone; 6, mesocone; 7, protocone.

Order Artiodactyla
Family Tayassuidae

Tayassuidae, gen. et sp. indet.
Fig. 38

REFERRED SPECIMENS. RdP_4 (NMC 10024), R. lower molar (SMNH P552), LdP^4 (SMNH P237).

DESCRIPTION

NMC 10024, RdP_4 (Fig. 38A), made up of six major cusps arranged in three transverse lophs; each loph made up of two major cusps on lingual and buccal edges, plus anterior and posterior accessory cuspules, elements of each loph arranged in lozenge; each lozenge with basin or basin system; basins of anterior and posterior lophs relatively simple, that of central loph complex.

Lower molar, SMNH P552 (Fig. 38B), badly broken; anterior accessory cuspule present; paraconid double; posterior accessory cuspule posterior to hypoconid.

LdP^4, SMNH P237 (Fig. 38C), with four major cusps, same accessory cusps and cingula as in tayassuid permanent molars; mesocone attached to hypocone by low loph; anterior accessory cuspule anterior to paraconid; anterior cingulum strong; strong posterior cingulum bearing cuspule, four crenulations; small accessory cuspules posterobuccal to protocone, anterior to metacone.

MEASUREMENTS

dP_4, AP(1), 13.3; BL(1), 7.6. dP^4, AP(1), 10.7; BL(1), 9.2.

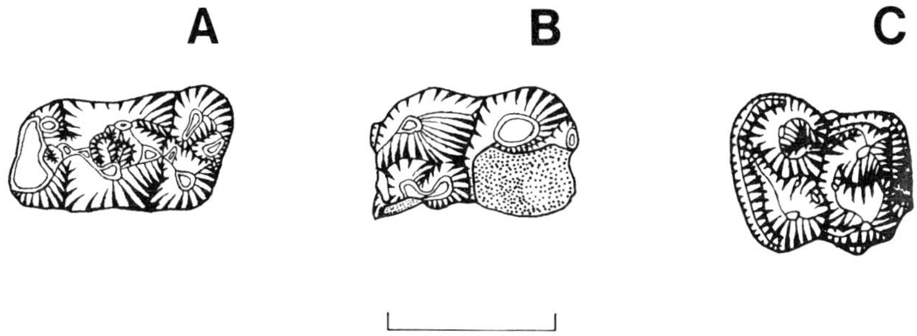

Fig. 38 Tayassuidae, gen. et sp. indet., cheek teeth in occlusal aspect. Anterior to left, buccal to top of page. Scale 1 cm.

A. RdP_4, NMC 10024.

B. R. lower molar, SMNH P552.

C. LdP^4, SMNH P237.

DISCUSSION

Reference to Lundelius' description and drawings (1960) has allowed me to assign these Wood Mountain specimens to their proper relative positions in the tooth rows. The tayassuid affinities of these specimens are clear, and two additional specimens in the present collections, an upper canine (NMC 10024) and an upper deciduous premolar (NMC 9442), may represent the same taxon but are not assigned because they must be compared to additional material.

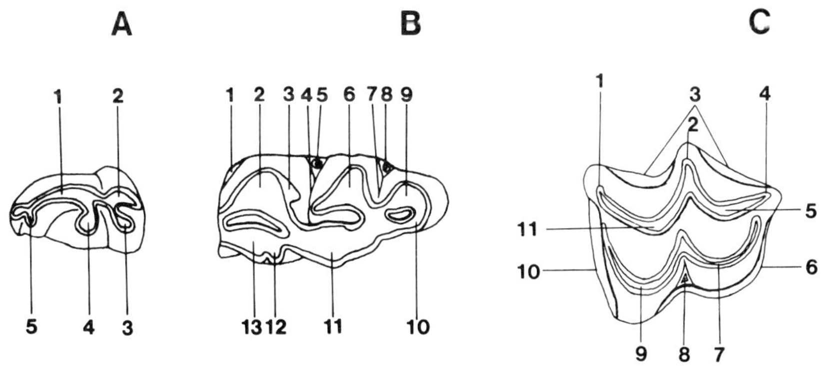

Fig. 39 Selenodont Artiodactyla. Terminology used in description of cheek teeth. Anterior to left, buccal to top of page. Not to scale.

A. RP_4: 1, protoconid; 2, hypoconid; 3, entoconid; 4, metaconid; 5, paraconid.

B. RM_3: 1, anterior cingulum; 2, protoconid; 3, *Palaeomeryx* fold; 4, protoflexid; 5, anterior ectostylid; 6, hypoconid; 7, hypoflexid; 8, posterior ectostylid; 9, hypoconulid; 10, entoconulid; 11, entoconid; 12, metastylid; 13, metaconid.

C. LM^1 or M^2: 1, parastyle; 2, mesostyle; 3, buccal cingula; 4, metastyle; 5, metacone; 6, posterior cingulum; 7, metaconule; 8, entostyle; 9, protocone; 10, anterior cingulum; 11, paracone.

Family Merycoidodontidae

***Ticholeptus* sp.**
Fig. 40

REFERRED SPECIMENS. L. mandibular fragment with alveoli for LI_{1-3}, broken LC_1, alveolus for LP_2, broken LP_3, full LP_4, LM_{1-3} (NMC 8982), RdP_4 (ROM 7696), L. lower molar (ROM 7693), L. upper molar (SMNH P384; ROM 7510).

DESCRIPTION

Three lower incisors in life, alveoli present in mandibular fragment, NMC 8982 (Fig. 40A, B); lower canine large, prominent, with nearly triangular cross section.

Lower cheek teeth mesodont; teeth preserved in NMC 8982, except for P_4 and M_3, broken or badly worn. P_3 badly broken; P_4 with short lingual valley, located just behind paraconid and protoconid, directed at about 70° to axis of tooth row, ending in small notch, not reflected by lingual striid; low cingulum posterior to hypoconid.

Weak buccal cingulum in molars; anterior ectostylids blocking protoflexids of M_2, M_3; M_3 with no posterior ectostylid, buccal cingulum not extending to enamel margin of hypoconulid.

Mandible (Fig. 40B) not deepening greatly at angle; mental process pronounced; two mental foramina present, one beneath anterior, one beneath posterior root of P_2; canine crowded back beneath roots of P_2; posterior end of P_2 displaced buccad at angle of about 45° to axis of tooth row.

Most nearly complete upper molar from Wood Mountain Formation, ROM 7510 (Fig. 40C), well preserved, but part of anterobuccal corner missing; oreodont characteristics including distinctive anterior, posterior, and buccal cingula; parastyle, mesostyle strong, metastyle weak.

MEASUREMENTS

P_3, AP(1), 11.0; BL(1), 7.6. P_4, AP(1), 13.8; BL(1), 9.3. M_1, AP(1), 12.1; BL(1), 9.7. M_2, AP(1), 15.5; BL(1), 12.2. M_3, AP(1), 24.6; BL(1), 12.0. Upper molar, AP(2), OR 21.4–22.1; BL(1), 19.7.

DISCUSSION

The ramal fragment, NMC 8982, conforms in development of the buccal cingulum and degree of hypsodonty of molars to that described for the genus *Ticholeptus*. This specimen cannot be assigned to any described species by its size, as most described species are similar in size. Five species and subspecies of *Ticholeptus* (data from Schultz and Falkenbach, 1941, p. 87) and NMC 8982 show the following anteroposterior diameters of molar series: *Ticholeptus hypsodus*, 56; *T. obliquidens*, 56; *T. tooheyi*, 52.5; *T. zygomaticus zygomaticus*, 55.5; *T. zygomaticus smithi*, 51.5; NMC 8982, 52.2.

Features of NMC 8982 which differ from previously described specimens assigned to *Ticholeptus* are marked displacement of P_2 and two mental foramina. These may well be individual characters and cannot be used to separate the Wood Mountain oreodont from any of the species listed above. The upper molar described above is of the right size to belong to the same taxon as NMC 8982 but otherwise lacks diagnostic generic characters.

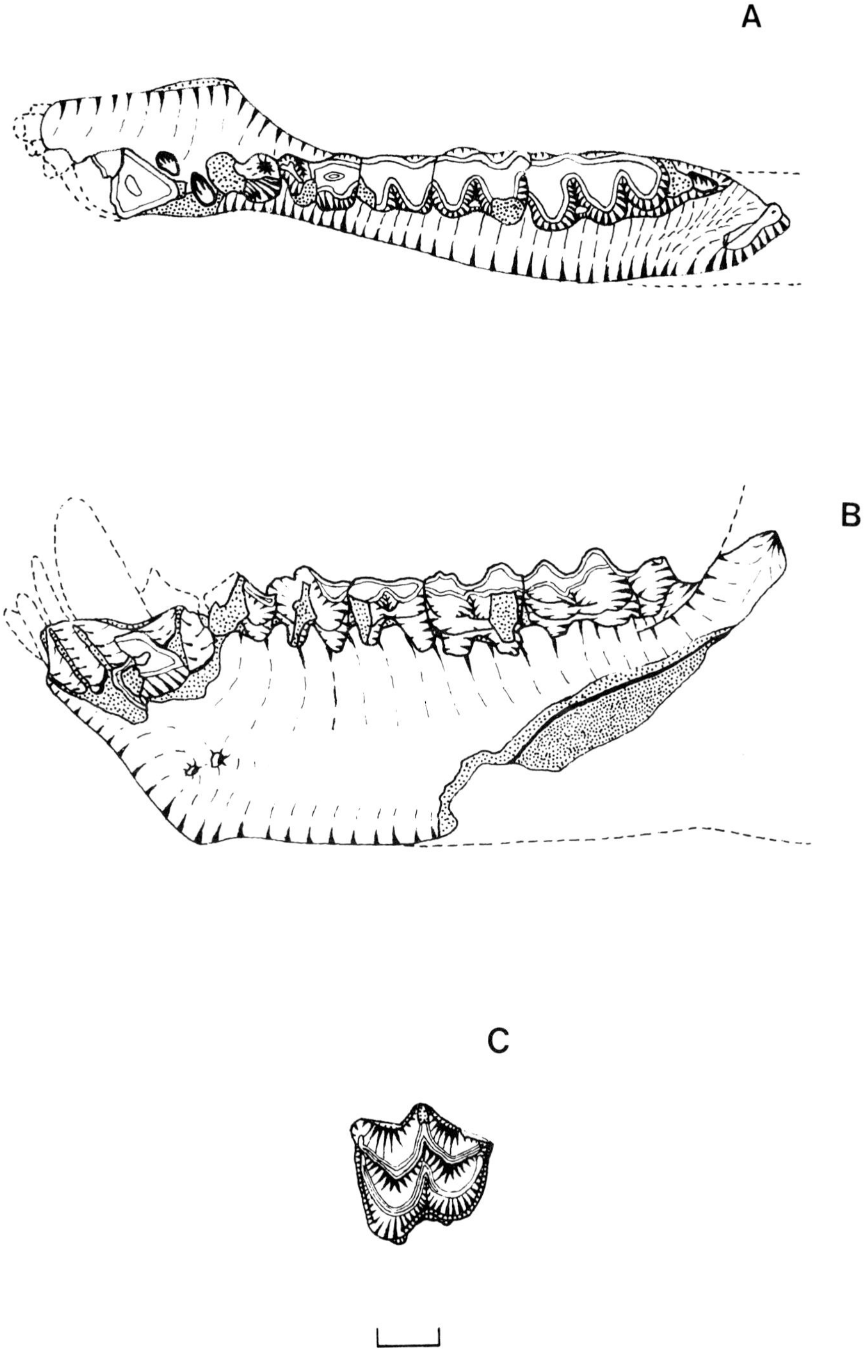

Fig. 40 *Ticholeptus* sp., mandible and cheek teeth. Anterior to left. Scale 1 cm.

A,B. L. mandibular fragment with LP_{3-4}, LM_{1-3}, NMC 8982.

A. Occlusal aspect, buccal to bottom of page.

B. Lateral aspect, occlusal to top of page.

C. L. upper molar, ROM 7510, occlusal aspect, buccal to top of page.

Family Camelidae

Procamelus sp.
Fig. 41

REFERRED SPECIMENS. LdP_4 (NMC 9033), R. mandibular fragment with broken RP_1, alveolus for RP_2, full RP_{3-4}, RM_{1-3} (NMC 9063), L. mandibular fragment with alveoli for LP_{2-3}, full LP_4, LM_{1-3} (SMNH P385), LM_2 (ROM 7485), LP^4 (NMC 9399), R. upper molar (NMC 8996, 8997; ROM 7486).

DESCRIPTION
Cheek teeth mesodont; P_1 (Fig. 41A, B) nearly oval in cross section, single-rooted; P_2 alveolus separated from that of P_1 by diastema.

Patton's description (1969, pp. 156–157) of referred specimens of *Procamelus occidentalis* exactly fitting known Wood Mountain premolars: "The two posterior premolars resemble one another closely in cusp morphology. They are marked by a pronounced anterolingual fold (flexid) between the anterior (paraconid?) and middle (metastylid?) lobes.... The posterior lobe of P_3 and P_4 is divided into two distinct cusps: a broad labial hypoconid and a narrow, less pronounced lingual entoconid. The entoconid attains much greater development in P_4 [than in P_3];" posterior premolars slightly reduced.

All known lower molars from Wood Mountain Formation well worn; no cingulum at enamel margin; no anterior, posterior ectostylids; lingual surfaces of crowns relatively smooth; metastylids of M_1 to M_3 weak.

Horizontal ramus (Fig. 41B) dorsoventrally slender; prominent mental process beneath P_1; single mental foramen posteroventral to P_1.

Isolated upper molars (Fig. 41C, D) typically camelid; no entostyle; no cingula at enamel margins; parastyle, paracone, and mesostyle protruding sharply from buccal edge of occlusal surface; metastyle weak.

DISCUSSION
Mandibles and cheek teeth of the genera *Protolabis* and *Procamelus* are so similar that it is difficult to distinguish them. Webb (1969, p. 148) revised the generic diagnosis of *Protolabis* and established criteria by which mandibles may be assigned to this genus; he made the following relevant points: "P_1 double rooted. Cheek teeth proportionally narrower than in *Procamelus*.... $P_{\overline{3}}$ and $P_{\overline{4}}$ with posterior cusps narrower than middle cusps and with much weaker lingual inflections of fossettes than in *Procamelus*." A classical distinction between *Procamelus* and *Protolabis* may be added to these: *Protolabis* shows greater reduction in size of P_3 and P_4 than does *Procamelus*.

These criteria suggest that the Wood Mountain camelid material should be assigned to *Procamelus* rather than to *Protolabis*. P_1 of NMC 9063 is single-rooted, the cheek teeth are proportionally as wide as in the specimens referred to *Procamelus occidentalis* by Patton (Table 9), P_3 is wider posteriorly than in the middle (Fig. 41A), and the posterolingual element of P_3 is somewhat lingually inflected.

Comparison of the ratio of the anteroposterior crown lengths of P_3 and M_2 indicates the degree of reduction of the posterior premolars. NMC 9063 has a ratio of 0.50, which is intermediate between that of Patton's specimen of *Procamelus occidentalis* (1969, pp. 165–166) at 0.58 and the specimen of *P. occidentalis* reported by McGrew (1938, p. 322) at 0.41. Alternatively, a specimen of *Protolabis angustidens* mentioned by Matthew (1901) has a ratio of 0.35. Thus the degree of reduction of the posterior premolars of NMC 9063 more closely resembles *Procamelus* than *Protolabis*.

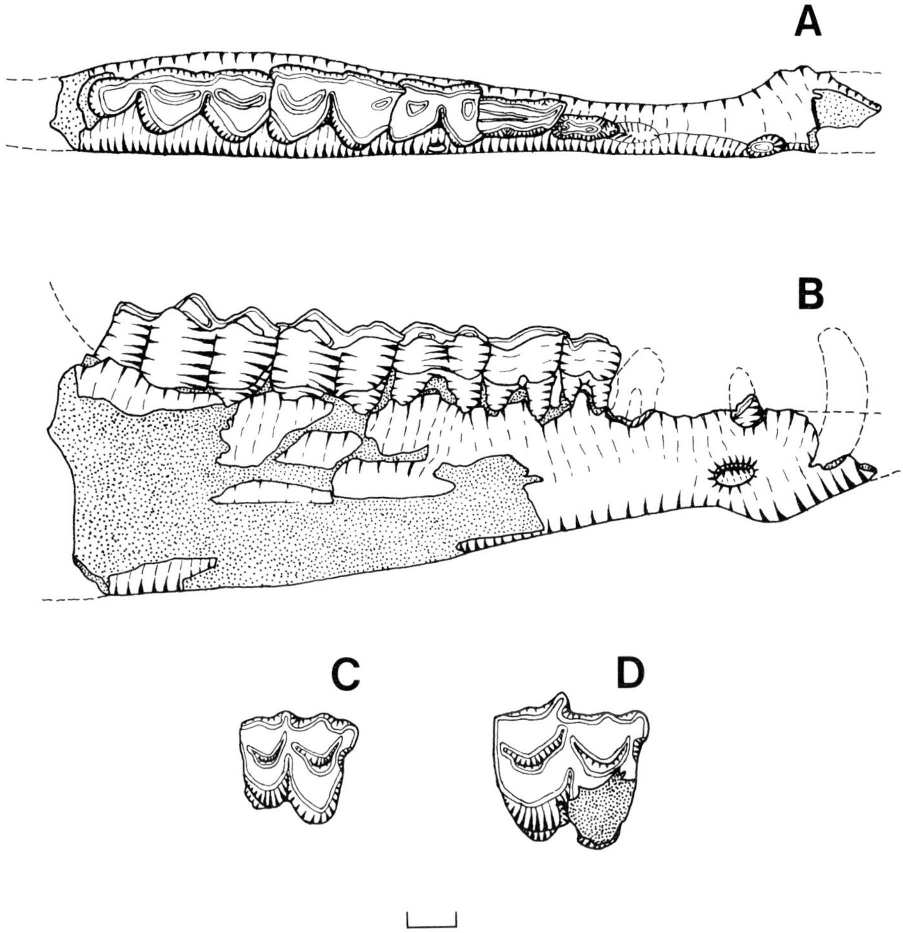

Fig. 41 *Procamelus* sp., mandible and cheek teeth. Anterior to right. Scale 1 cm.
A,B. R. mandible with RP_1, RP_{3-4}, RM_{1-3}, NMC 9063.
A. Occlusal aspect, buccal to bottom of page.
B. Lateral aspect, occlusal to top of page.
C. R. upper molar, NMC 8997, occlusal aspect, buccal to top of page.
D. R. upper molar, NMC 8996, occlusal aspect, buccal to top of page.

The Wood Mountain camelid differs from Patton's specimens of *Procamelus occidentalis* (1969, fig. 16) in the position of the mental foramen, which is set anteroventral to P_1 in the Texas Gulf Coast specimens, and in the length of the diastema separating P_1 and P_2, as well as in the depth of the ramus beneath the diastema. It is comparable to *P. occidentalis* in size and probably belongs to the *P. occidentalis* group (Patton, 1969, p. 155).

Table 9. Measurements of teeth and mandible of *Procamelus* sp. from Wood Mountain compared with Patton's measurements of *P. occidentalis* (1969, table 12).

Tooth	Dimension	n	OR Wood Mountain	OR Patton measurements (n = 2 or 3)
P^4	AP	1	15.6	—
	BL	1	13.7	—
Upper molars	AP	3	24.3–31.3	—
	BL	1	20.7	—
P_1	AP	1	7.0	—
	BL	1	3.6	—
P_3	AP	1	13.1	12.7–13.5
	BL	1	5.3	5.8–6.0
dP_4	AP	1	28.4	—
	BL	1	10.7	—
P_4	AP	2	13.0–17.0	14.0–15.9
	BL	2	6.5–7.7	7.4–8.5
M_1	AP	2	15.8–18.4	20.1–21.0
	BL	2	11.6–12.8	12.0–12.5
M_2	AP	2	22.5–26.1	23.7–26.0
	BL	3	14.9–16.0	12.3–15.0
M_3	AP	2	34.7–35.4	34.6–38.1
	BL	2	12.7–14.9	13.8–15.0
Mandible				
Diastema, P_1-P_2,	AP	2	15.8–22.6	30.1
DV, ramus beneath centre of diastema		2	18.2–19.5	24.2–27.6
DV, ramus beneath centre of M_1		2	31.0–35.0	30.0–35.2

Family Palaeomerycidae

Blastomeryx cf. *B. elegans* Matthew and Cook, 1909
Fig. 42

REFERRED SPECIMENS. LdP_4 (ROM 7695), RP_3 or P_4 (ROM 7500), LM_1 or M_2 (NMC 9442—2 specimens; ROM 7499, 7501, 7511), RM_1 or M_2 (SMNH P236, P557, P558; NMC 9442; ROM 7498, 7504–7506), R. mandibular fragment with RM_1, broken RM_2 (NMC 9027), LM_3 (SMNH P551, 8974; NMC 9442—2 specimens), RM_3 (SMNH P122), L. mandibular fragment with LM_{2-3} (SMNH P237), C^1 (NMC 8966, 9442), L. upper molar (SMNH P123, P551, P566; NMC 9026; ROM 7508, 7509), R. upper molar (SMNH P552; NMC 9440, 9442; ROM 7507).

DESCRIPTION

Teeth moderately hypsodont (Fig. 42A). Lower molars (Fig. 42B, C) without *Palaeomeryx* fold; anterior cingula present; metastylids not visible; hypoconulids persistently separated from hypoconids; entoconids strongly connected to metaconids, persistently separate from entoconulids. Upper molars (Fig. 42D) with strong, persistent buccal styles, prominent entostyles.

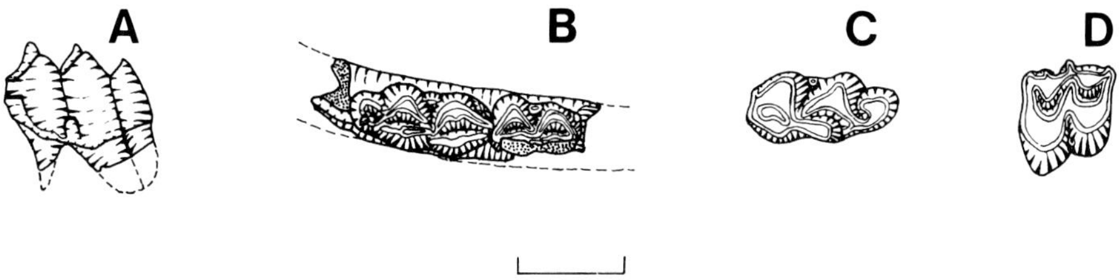

Fig. 42 *Blastomeryx* cf. *B. elegans*, cheek teeth. Scale 1 cm.
A. LM_3, NMC 9442, lateral aspect, anterior to left, occlusal to top of page.
B. LM_{2-3}, SMNH P237, occlusal aspect, anterior to right, buccal to top of page.
C. RM_3, SMNH P122, occlusal aspect, anterior to left, buccal to top of page.
D. L. upper molar, NMC 9026, occlusal aspect, anterior to left, buccal to top of page.

DISCUSSION

Upper premolars of *Blastomeryx* cf. *B. elegans* could not be distinguished from those of *Merycodus*. Upper molars are poorly preserved, and no dependable method of separating M^3 from M^1 or M^2 was found. Canines and molars are similar to those in the genus *Blastomeryx*. Elements available are poorly preserved but appear reasonably similar in size (Table 10), height of crown of cheek teeth, and degree of development of ectostylids and entostyles on molars, with those of the Lower Snake Creek *Blastomeryx elegans*. But M_3s from the Wood Mountain Formation are somewhat smaller than M_3 of the type. An upper canine, NMC 8966 (not shown), agrees roughly in size and shape, with the upper canine of *B. elegans*.

Table 10. Measurements of teeth of *Blastomeryx* cf. *B. elegans* compared with those of holotype of *B. elegans* (from Matthew and Cook, 1909, p. 410).

Tooth	Dimension	n	Mean	OR	CV	OR Holotype of *B. elegans*, AMNH 14101
dP_4	AP	1		12.0		
	BL	1		6.5		
P_3 or P_4	AP	1		5.8		6.2–7.3
	BL	1		2.9		3.0–4.0
M_1 or M_2	AP	13	8.3 ± 0.3	6.9–9.6	10.8	9.0–10.3
	BL	11	5.4 ± 0.2	4.4–6.3	13.0	5.2–5.9
M_3	AP	6	8.3 ± 0.1	8.2–8.5	1.2	13.9
	BL	6	5.9 ± 0.1	5.7–6.1	1.1	5.9
C^1	AP	2		6.0–8.2		
	BL	2		4.0–4.1		
Upper molar	AP	9	8.4 ± 0.3	7.0–9.6	10.7	
	BL	3	9.2	7.8–10.2		

***Cranioceras (Procranioceras)* cf. *C. (P.) skinneri* Frick, 1937**
Figs. 43, 44

REFERRED SPECIMENS. LdP_3 (SMNH P558), RP_3 (SMNH P566), LdP_4 (SMNH P550), LP_4 (SMNH P558), R. mandibular fragment with RdP_4, RM_1 (SMNH

P384), RM_2 (SMNH P122, P384), R. mandibular fragment with RM_{2-3} (NMC 9005), RM_{2-3} (ROM 7497), LM_3 (SMNH P566, 7993; ROM 7496), RM_3 (ROM 7495), RP^2 (NMC 9035), LM^1 or M^2 (SMNH P382, P552; ROM 7515), RM^1 or M^2 (NMC 9044, 9446), L. postorbital horn (NMC 8986).

DESCRIPTION

Teeth brachydont (Fig. 43A, B). Lower molars with *Palaeomeryx* fold, more prominent on M_1 and M_2 than on M_3; prominent anterior ectostylids on lower molars; small posterior ectostylid on M_3; lower molars with strong metastylids, prominent anterior cingula.

Upper molars (Fig. 44A) as those of *Blastomeryx* (Fig. 42D) in shape, pattern,

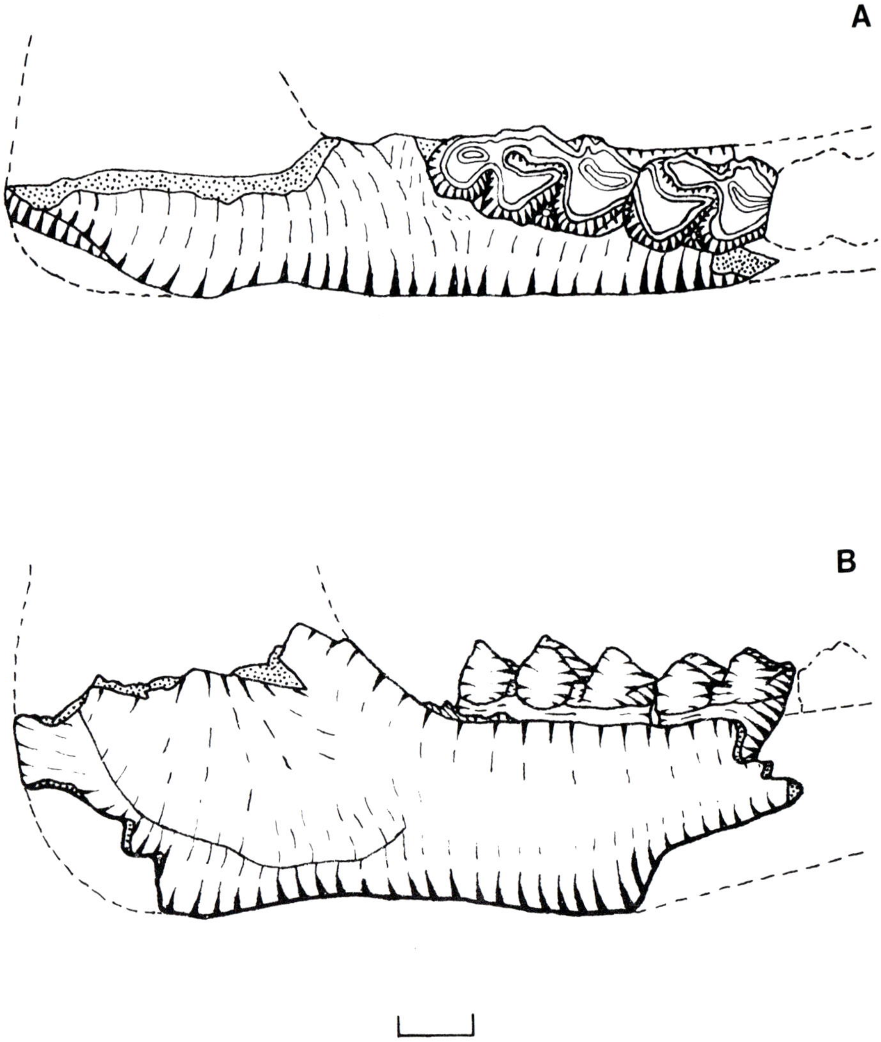

Fig. 43 *Cranioceras* (*Procranioceras*) cf. *C.* (*P.*) *skinneri*, R. mandibular fragment with RM_{2-3}, NMC 9005. Anterior to right. Scale 1 cm.
A. Occlusal aspect, buccal to bottom of page.
B. Lateral aspect, occlusal to top of page.

and presence of entostyle; both M^1 and M^2 represented, as indicated by high coefficients of variation below; M^3 could also be present.

Postorbital horn (Fig. 44B) heavy, long, lacking tip; strong postorbital bar, and posteromedial to it, strong flange leading to skull roof; supraorbital ridge heavy; cross sections (Fig. 44B) with the following shapes: basal cross section elliptical, major axis oriented anteroexternally, cross section taken 42 mm higher on shaft externally flattened oval; horn showing shallow anteroexternal groove only at base.

MEASUREMENTS

dP_3, AP(1), 15.6; BL(1), 10.0. P_3, AP(1), 13.6; BL(1), 6.9. dP_4, AP(2), OR 18.4–18.7; BL(2), OR 9.2–10.0. M_1, AP(1), 16.5; BL(1), 12.5. M_2, AP(3), mean 18.3, OR 17.5–19.1; BL(3), mean 13.8, OR 13.2–14.9. M_3, AP(5), mean 24.8 ± .9, OR 21.6–27.2, CV 8.5; BL(5), mean 12.8 ± .4, OR 11.3–14.1, CV

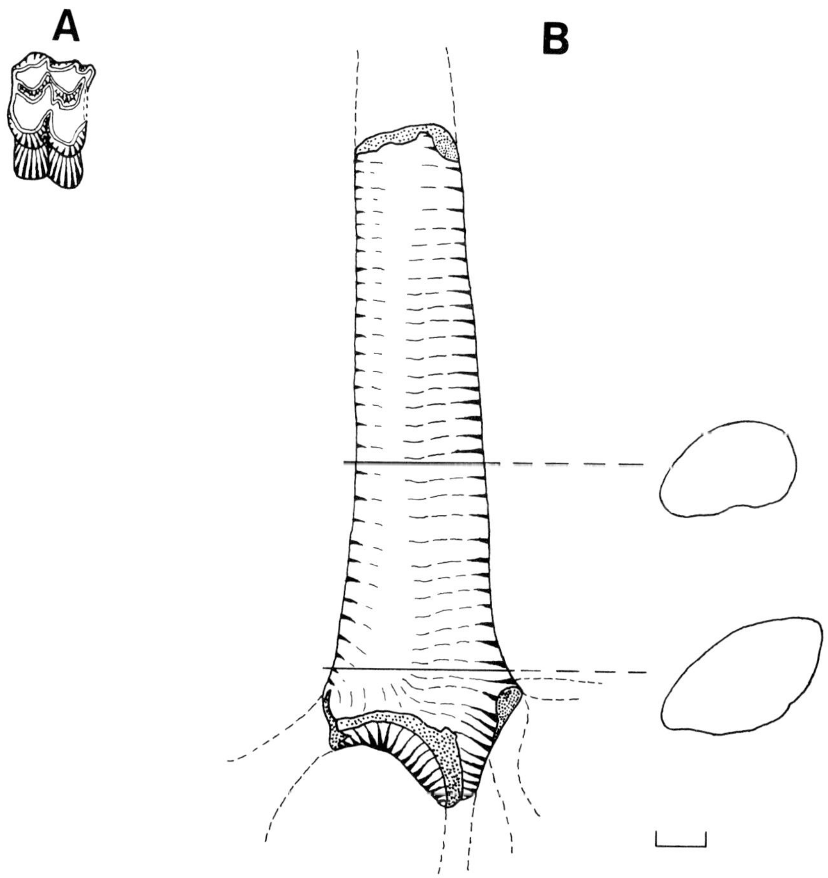

Fig. 44 *Cranioceras* (*Procranioceras*) cf. *C.* (*P.*) *skinneri*, tooth, horn. Scale 1 cm.

A. RM^1 or M^2, NMC 9044, occlusal aspect, anterior to right, buccal to top of page.

B. L. postorbital horn, NMC 8986, lateral aspect, anterior to left, dorsal to top of page. Cross-sections from levels indicated, rotated 90° on an axis parallel to page, external side downward.

7.0. P^2, AP(1), 14.4; BL(1), 7.1. M^1 or M^2, AP(5), mean 16.3 ± .8, OR 14.8–19.2, CV 10.4; BL(5), mean 19.0 ± 1.1, OR 16.0–23.5, CV 13.2.

DISCUSSION

This horn corresponds in cross section to that of the holotypic skull of *Cranioceras* (*Procranioceras*) *skinneri* (FAM 31250; *contra* Frick, 1937, fig. 12). Upper molars also resemble those of FAM 31250 in size and height of crown. This combination of characteristics suggests the identity of the Wood Mountain *Cranioceras* as *Cranioceras skinneri*. Only a relatively complete skull can confirm this identity. The characteristics of the lower jaw and teeth agree well with referred specimens of *C. skinneri*.

cf. *Sinclairomeryx* sp.

Fig. 45

REFERRED SPECIMEN. R. postorbital horn (ROM 7697).

DESCRIPTION

Horn slender, gently curved anterad; tip directed outward (Fig. 45B), the two fossae being dorsal wall of right orbit and dorsal right wall of braincase; surface covered with reticulate pattern of shallow grooves (not shown).

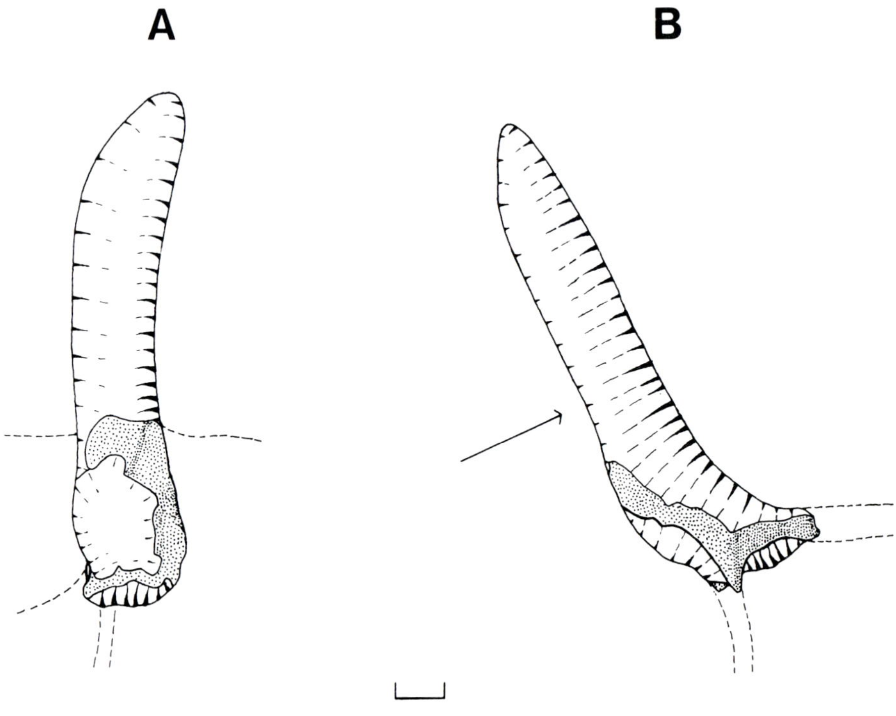

Fig. 45 cf. *Sinclairomeryx* sp., R. postorbital horn, ROM 7697. Dorsal to top of page. Scale 1 cm.

A. Lateral aspect, anterior to right.

B. Anterior aspect, lateral to left. Arrow indicates angle from which A was drawn.

DISCUSSION

The slender, anteriorly curved horn with the orientation shown in Fig. 45B shows slight resemblance to the corresponding postorbital horn of *Sinclairomeryx sinclairi* from the Sheep Creek Beds (Frick, 1937, fig. 16B). No other described taxon shows as great a similarity to ROM 7697.

The reticulate pattern of shallow grooves described above may indicate the paths of blood vessels beneath a layer of skin during life. Three possibilities are present; the taxon features short horns covered with skin, like the modern giraffe; or the individual was young, and the bony horns were still in a period of active growth throughout their length; or a deciduous sheath like that of *Antilocapra* was present.

Family Antilocapridae

Submeryceros minor serpentinus Frick, 1937

Fig. 46A, B

REFERRED SPECIMENS. LM_3 (NMC 9441), RM_3 (SMNH P122).

DESCRIPTION

Third lower molars moderately hypsodont, relatively small; last lower molar typically antilocaprid in pattern, with entoconulid and hypoconulid not prominent, but making up rounded posterior ridge enclosing deep valley, each cusp strongly connected to all neighbouring cusps at midline; strong anterior ectostylid blocks protoflexid; hypoflexid open; metastylid absent; cusps not projecting prominently linguad.

DISCUSSION

Both specimens match the drawing (Frick, 1937, fig. 44) of the holotype of *Submeryceros minor serpentinus*, FAM 31577. SMNH P122 matches the M_3 of FAM 31577 almost exactly in height of crown, length, and width (Table 11), and has a slightly larger anterior ectostylid. NMC 9441 is slightly larger but compares in height of crown and strength of the anterior ectostylid.

Additional specimens of *Submeryceros* are probably present in the collections from the Wood Mountain Formation. Isolated teeth other than M_3 of *Submeryceros* are extremely difficult to identify, as P_4, M_1, and M_2 are not appreciably smaller in size in *Submeryceros* than in *Merycodus necatus* or *M. sabulonis* and are only slightly shorter crowned. Thus only third lower molars have been assigned to *Submeryceros*.

Table 11. Measurements of teeth of *Submeryceros minor serpentinus* from Wood Mountain compared with those of holotype (estimated from Frick, 1937, fig. 44).

Tooth	Dimension	n	OR	Holotype of *S. minor serpentinus* FAM 31577
M_3	AP	2	9.2–11.0	11
	BL	2	4.1–4.8	5

Merycodus necatus Leidy, 1854

Fig. 46C–E

REFERRED SPECIMENS. L. horn cores (NMC 9065, 9409; ROM 7489, 7490, 7491, 7512, 7513), R. horn cores (SMNH 8874; NMC 9007—2 specimens, 9410—2 specimens, 9382; ROM 7514), fragmentary horn cores, probably of *M. necatus* (NMC 9007—9 specimens, 9065—2 specimens, 9410; ROM 7488, 7493, 7494).

DESCRIPTION

Horn cores relatively short; most with characteristic branching into small, procumbent anterior tine and larger, erect posterior tine; juvenile specimen (Fig. 46C) with two tines of about equal size; horn core with longest shaft (NMC 9007; Fig. 46E) 4 cm from dorsal wall of orbit to notch between tines; much of variability in shape of horn core related to age of individual, different in young (Fig. 46C), mature (Fig. 46D), old animals (Fig. 46E); if burr present, appearing just below increase in diameter beneath tines.

DISCUSSION

In discussions of antilocaprids, I refer to the specimens representing postorbital horns as "horn cores", because of several important resemblances to the horn cores of modern *Antilocapra*. As Matthew (1924, pp. 202–206) stated, "The burr ... may be single, double, or triple, in two cases quadruple. It is very easily broken off and leaves no perceptible scar.... The beam is NEVER FOUND BROKEN OFF AT THE BURR.... Obviously a multiple burr could not normally occur with a deciduous antler. But in *Merycodus* the multiple burr is certainly a normal character. It appears to me far more reasonable, therefore, to conclude that the bony cores were not deciduous ... but their sheath periodically renewed as in *Antilocapra.*"

Russell (1937, pp. 78–80) described NMC 8874, a right horn core (there designated GSC No. 8774) and pointed out its resemblance to *Merycodus necatus*. As he stated, "... the burr in the Saskatchewan specimen is not complete, but is divided into three portions, one small and two large. The intervening areas between these parts show faint grooves and ridges.... During the early part of the animal's life the antler was covered with skin, sufficiently vascularized to nourish the growing bone. When full growth had been reached the skin on the upper

Fig. 46 Antilocapridae, lower molars and horn cores. Scale 1 cm.
- A,B. *Submeryceros minor serpentinus*, RM_3, SMNH P122, anterior to right.
- A. Lateral aspect, occlusal to top of page.
- B. Occlusal aspect, buccal to bottom of page.
- C-E. *Merycodus necatus*, horn cores, lateral aspects, dorsal to top of page.
- C. R., juvenile, ROM 7514, anterior to right.
- D. L., NMC 9409, anterior to left.
- E. L., NMC 9007, anterior to left.
- F. *Merycodus sabulonis*, L. horn core, ROM 7487, lateral aspect, dorsal to top of page, anterior to left.
- G. *Merycodus necatus* or *M. sabulonis*, LM_3, NMC 9400, lateral aspect, anterior to left, occlusal to top of page.
- H,I. *Cosoryx* (*Paracosoryx*) cf. *C.* (*P.*) *alticornis*.
- H. L. mandibular fragment with LM_3, ROM 7492, lateral aspect, anterior to left, occlusal to top of page.
- I. ?L. horn core, SMNH P237, lateral aspect, anterior to left, dorsal to top of page.

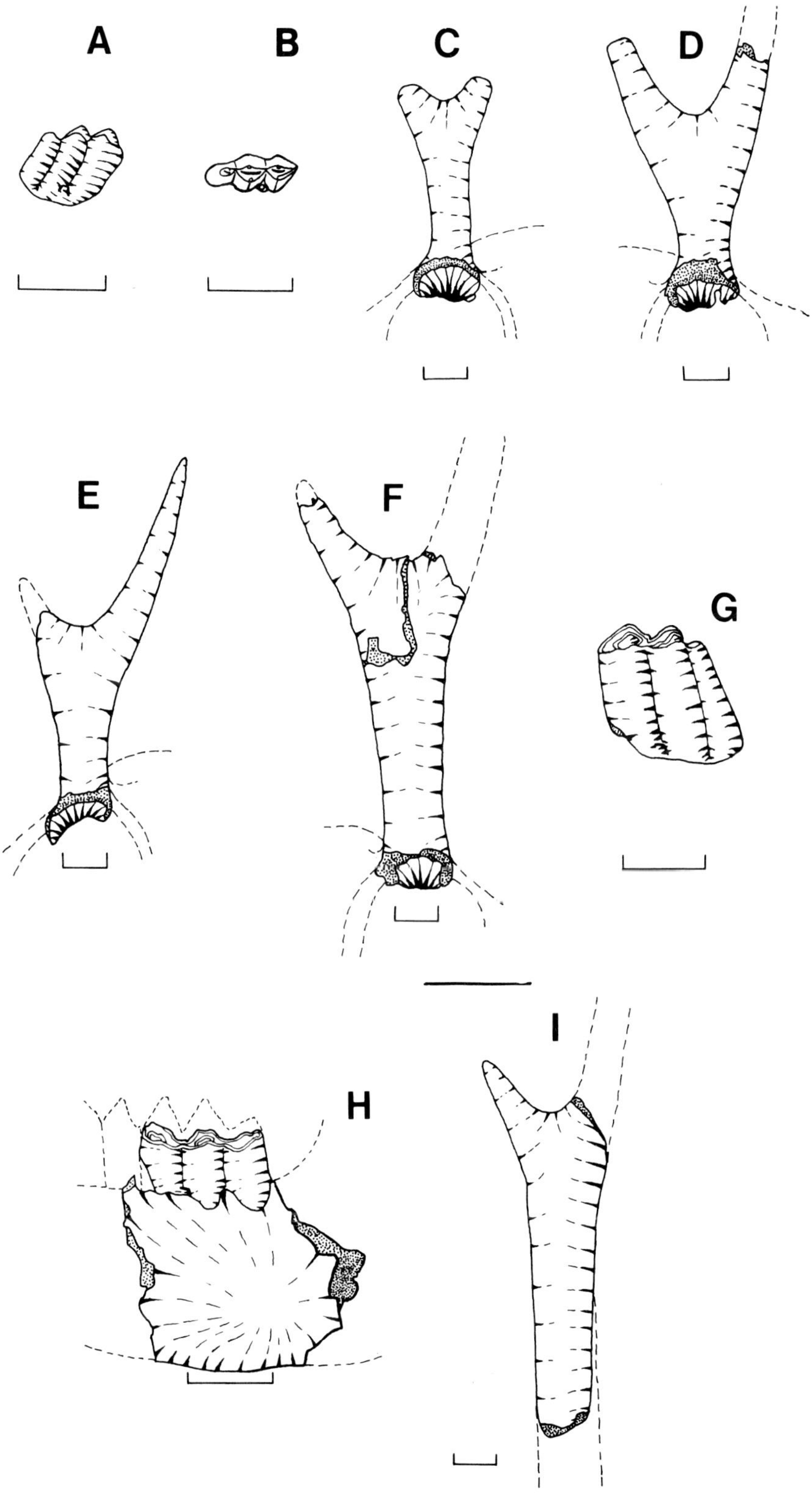
A
B
C
D
E
F
G
H
I

portion might tend to degenerate, as it would be a useless or even objectionable covering for a structure used in combat. With this degeneration a burr might develop at the upper margin of the still living skin.... The Wood Mountain specimen appears to represent a stage when the skin of the upper portion was not yet wholly degenerate, as there is still space for vascular connections with the basal portion." Corroborative evidence was given by Voorhies (1969, p. 40), in connection with some Lower Pliocene specimens from Nebraska, stating, "That the skin was absent, at least from the tips, after horn growth ceased is indicated by the fact that most tips show polishing as though they had been worn during use."

Horn cores assigned to *Merycodus necatus* have not been measured both because of their relative incompleteness and because of proportional changes that occur with the aging of the animal. Skinner and Taylor stated (1967, p. 41) that "the horn shaft of *M. necatus* is shorter and the prongs longer than in *M. sabulonis*. In both *M. necatus* and *M. sabulonis*, the burr is centrally placed on the shaft and is relatively higher than that in *Cosoryx furcatus* and *Meryceros warreni*...." Specimens from the Wood Mountain Formation are assigned to *M. necatus* and *M. sabulonis* in agreement with Skinner and Taylor's statement.

No teeth or jaws have been specifically assigned to either *M. necatus* or *M. sabulonis*. The teeth of both are hypsodont, and no consistent morphological feature is known by which they can be separated.

Merycodus sabulonis **Matthew and Cook, 1909**
Fig. 46F

REFERRED SPECIMENS. L. horn cores (SMNH P391, P562; NMC 9410(?); ROM 7487), R. horn cores (SMNH P391, P552, P566).

DESCRIPTION

Horn cores longer-shafted than in *M. necatus*, shortest measuring 5.5 cm from dorsal edge of orbit to notch between tines, apparently from a juvenile animal; typical adult specimen (Fig. 46F) nearly 7 cm; burrs present on two specimens, midway along shaft.

DISCUSSION

These few specimens, mainly obtained from localities south of Fir Mountain, correspond to the drawings of Skinner and Taylor (1967, fig. 11) and to specimens assigned to *M. sabulonis* in the Frick Collection, American Museum of Natural History. They are distinguished from horn cores of *M. necatus* in having longer shafts. Dental material has not been assigned specifically to *M. sabulonis* but is discussed under a separate heading.

Merycodus necatus **or** ***M. sabulonis***
Fig. 46G

REFERRED SPECIMENS. Numerous teeth and jaw fragments found at nearly all localities in the Wood Mountain Formation (Storer, 1970*b*, p. 189).

DESCRIPTION

Teeth hypsodont, bearing same typically antilocaprid pattern as in *Submeryceros* (Fig. 46B); anterior ectostylids strong; hypoflexids in M_3 open; posterior column, bearing hypoconulid and entoconulid in M_3, shorter anteroposteriorly than anterior and median columns, even at base of crown.

DISCUSSION

The specimens are referable to *Merycodus*, but it is not possible to distinguish dental material of *M. necatus* and *M. sabulonis*.

M_3 is easily distinguished from the corresponding teeth of the Wood Mountain *Submeryceros* and *Cosoryx* on the base of overall size and proportional length of the posterior column. Other cheek teeth are typically merycodont.

Cosoryx (Paracosoryx) cf. *C. (P.) alticornis* Frick, 1937

Fig. 46H, I

REFERRED SPECIMENS. L. mandibular fragment with LM_3 (ROM 7492), RM_3 (ROM 7694), ? L. horn core (SMNH P237).

DESCRIPTION

Teeth (Fig. 46H) large, hypsodont; occlusal pattern similar to those of *Submeryceros* and *Merycodus* (Fig. 46B, G); anterior ectostylid strong; hypoflexid open; posterior column bearing entoconulid and hypoconulid as long anteroposteriorly as anterior and median columns.

Horn core (Fig. 46I) lacking base, but apparently with very long shaft; erect posterior tine apparently larger than procumbent anterior tine.

MEASUREMENTS

M_3, AP(1), 20.2; BL(2), OR 5.9–7.4.

DISCUSSION

The horn core compares favorably with those illustrated by Frick (1937, fig. 308A) for *Cosoryx (Paracosoryx) alticornis*, and with specimens 31152 and 31154 in the American Museum of Natural History. M_3s can also be referred to this species on similar size and degree of hypsodonty.

M_3 differs from corresponding teeth of *Merycodus necatus* and *M. sabulonis* in its larger size and the great anteroposterior length of the posterior column, which is parallel-sided. This latter factor may not be of great importance, however, as Skinner and Taylor (1967, fig. 9C) recorded a parallel-sided posterior column in an M_3 of *Cosoryx furcatus*.

Order Rodentia
Family Mylagaulidae

Mylagaulus cf. *M. laevis* Matthew, 1902

Figs. 48, 49, 50

REFERRED SPECIMENS. R. mandibular fragment with RI_1, RdP_4 (NMC 8969), R. mandibular fragment with RI_1, RP_4 (NMC 8968), R. mandibular fragment with RI_1 (SMNH P123—2 specimens), LI_1 (SMNH P123—2 specimens), LdP_4 (ROM 7278), LP_4 (SMNH P122—2 specimens, 7990; NMC 9406; ROM 7279), RP_4 (SMNH P122—3 specimens, P123; NMC 8970; ROM 7280), LM_1 or M_2 (ROM 7281, 7282), RM_1 or M_2 (ROM 7283–7286), L. mandibular fragment with LP_4, LM_{2-3} (SMNH (?)—see under "Collections" in Materials and Methods), LdP^4 (ROM 7287), RdP^4 (ROM 7288), LP^4 (NMC 9432; ROM 7291, 7292), RP^4 (SMNH P122—2 specimens, P123, P391; NMC 8971, 9406, 9430, 10026; ROM 7289, 7290), LM^1 or M^2 (ROM 7293), RM^1 or M^2 (ROM 7294), RM^3 (ROM 7295, 7296), ?M^3 (ROM 7297).

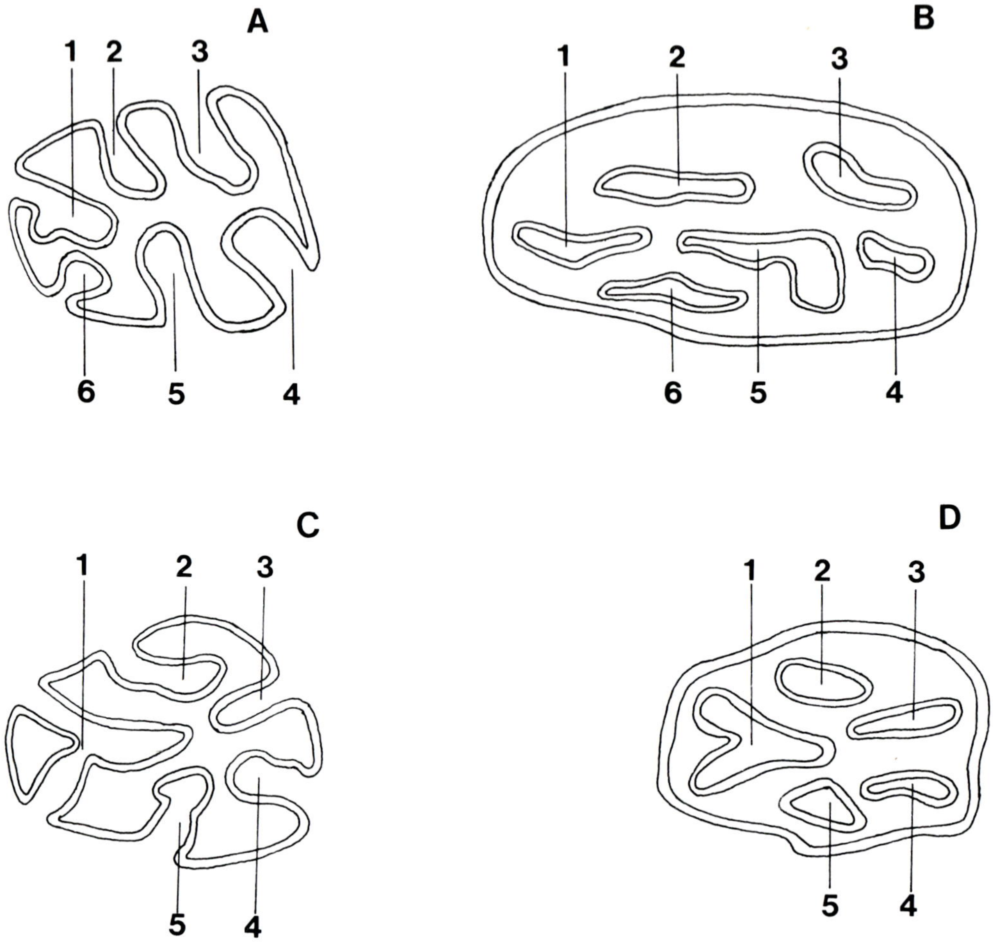

Fig. 47 *Mylagaulus* cf. *M. laevis*. Terminology used in description of premolars. Anterior to left; buccal to top of page in A,B, to bottom in C,D. Not to scale.

A. RP_4 lightly worn with flexids open: 1, anteroflexid; 2, protoflexid; 3, hypoflexid; 4, entoflexid; 5, mesoflexid; 6, metaflexid.

B. RP_4 more heavily worn with flexids closed: 1, anterofossettid; 2, protofossettid; 3, hypofossettid; 4, entofossettid; 5, mesofossettid; 6, metafossettid.

C. RP^4 lightly worn with flexi open: 1, paraflexus; 2, protoflexus; 3, hypoflexus; 4, metaflexus; 5, mesoflexus.

D. RP^4 more heavily worn with flexi closed: 1, parafossette; 2, protofossette; 3, hypofossette; 4, metafossette; 5, mesofossette.

DESCRIPTION

Lower and upper deciduous premolars and molars poorly preserved, not described; similar to corresponding elements of other described species of *Mylagaulus*; molars determined following criteria of Black and Wood (1956).

Anterofossettid of P_4 forked as formed by closing of anteroflexid (Fig. 48A); forked condition persisting (Fig. 48B) as other flexids close, or disappearing

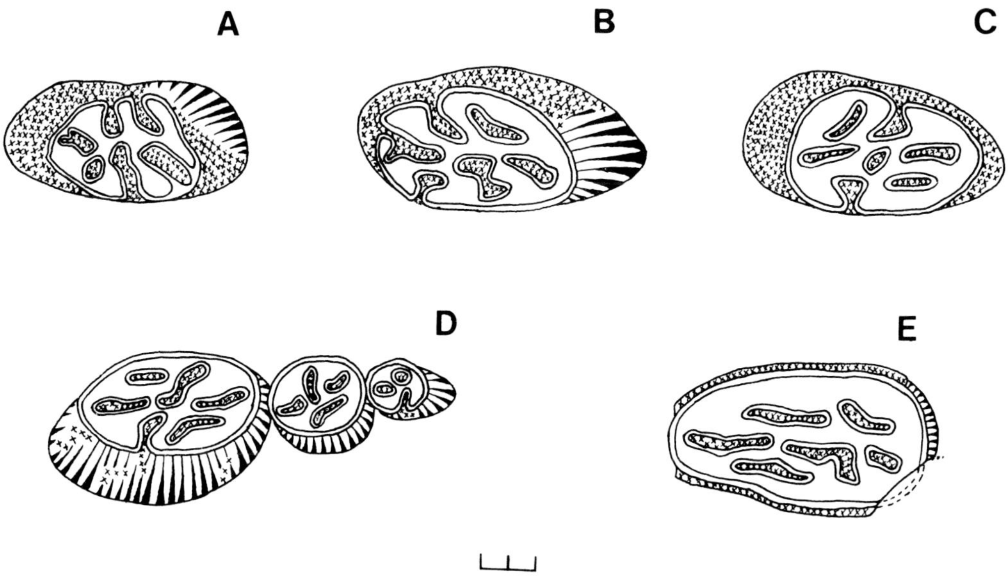

Fig. 48 *Mylagaulus* cf. *M. laevis*, lower premolars in occlusal aspect. Scale in mm.
A. RP_4, SMNH P122, anterior to left, buccal to top of page.
B. RP_4, SMNH P123, anterior to left, buccal to top of page.
C. LP_4, NMC 9406, anterior to right, buccal to top of page.
D. LP_4, LM_{2-3}, SMNH(?), anterior to left, buccal to bottom of page.
E. RP_4, NMC 8968, anterior to left, buccal to top of page.

(Fig. 48C); protoflexid last to close (Figs. 48D, 49A), completely closed only in well-worn specimens (Fig. 48E); all flexids filled with cement, forming cement-filled fossettids in closing; mesofossettid persisting as single S-shaped basin in most specimens (best shown in Figs. 48B, D; 49A), may split into two sections on closing of mesoflexid (Fig. 48C), but most similarly worn specimens with continuous mesofossettid (Figs. 48D, E; 49A); seven fossettids in some specimens, most worn lower premolars with six fossettids.

Lower jaws resembling type material of *Mylagaulus laevis*. In young individual, mandibular angle well represented (Fig. 49B); ascending ramus long, stout, arising just posterior to P_4; masseteric ridge strong; mental foramen high on jaw just anterior to P_4. In NMC 8968, adult individual (not shown), masseteric ridge far more prominent than in young individual; entire ramus far deeper in NMC 8968; greater space between dorsal surface of ramus and mental foramen.

In P^4, mesoflexus and protoflexus last to close (Fig. 50A); this specimen with forked parafossette, double metafossette; mesoflexus in shape of elongate S; six fossettes and flexi present; flexi filled with cement.

In more worn specimens, two patterns of wear appearing; in ROM 7289 (Fig. 50B), both mesofossette and metafossette double, but mesofossette and metafossette joined; in SMNH P122 (Fig. 50C), mesofossette and metafossette separate, neither double; ROM 7289 like other upper premolars figured (Fig. 50D, E), having doubled mesofosette and metafossette.

In specimens representing old individuals (Fig. 50D), forked parafossette split into short lingual fossette, elongate buccal fossette; mesofossette and hypofossette elongated, mesofossette and metafossette double.

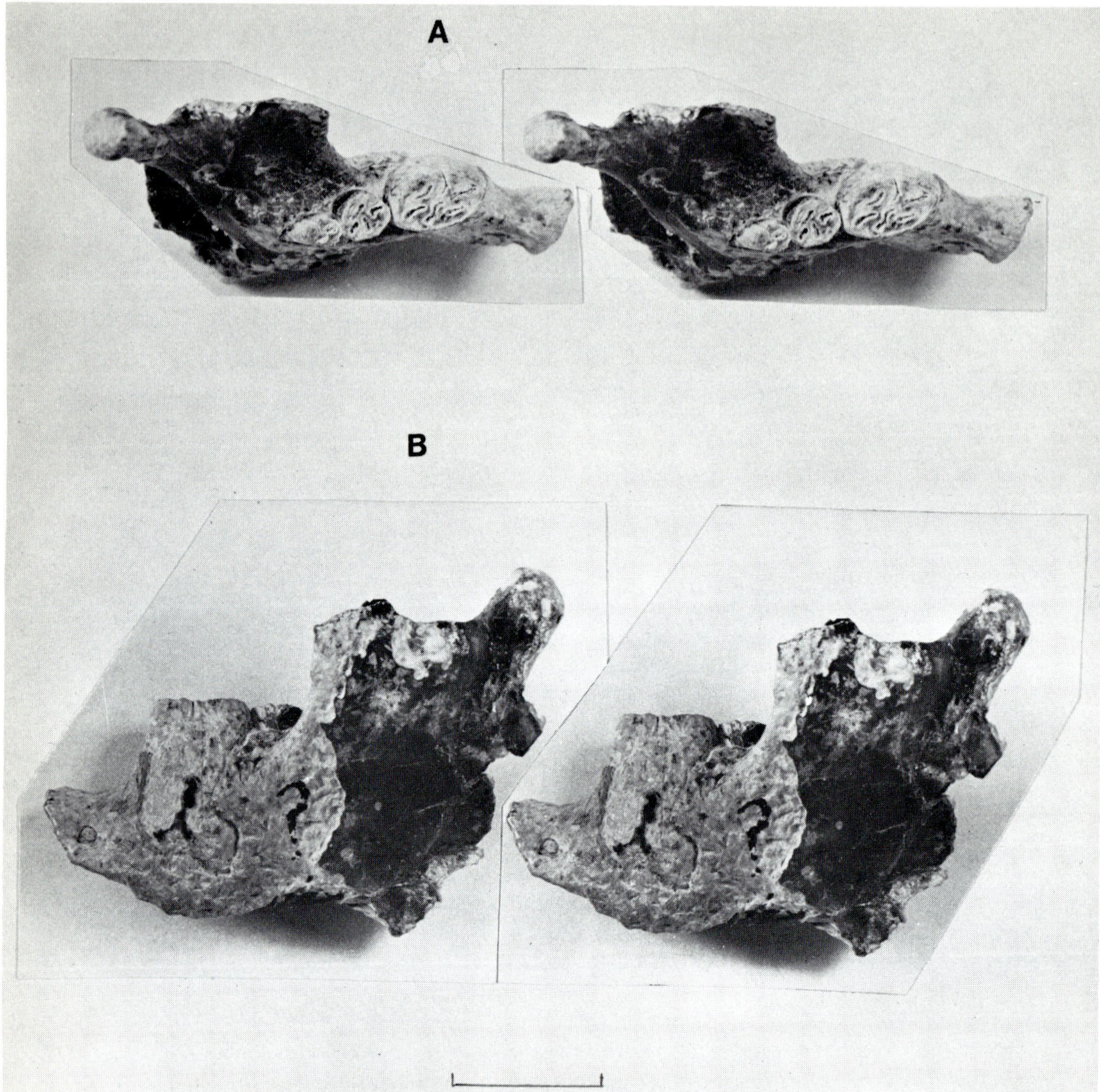

Fig. 49 *Mylagaulus* cf. *M. laevis*, L. mandible with LP_4, LM_{2-3}, SMNH(?). Scale 1 cm.
A. Occlusal aspect, anterior to right, buccal to top of page.
B. Lateral aspect, anterior to left, occlusal to top of page.

In NMC 9432 (Fig. 50E), parafossette split into two short anterior fossettes and single elongate posterior fossette; no other major changes from condition seen in NMC 10026; tooth with nine fossettes, resembling late wear stages of most species of *Mylagaulus*, especially *M. monodon*.

MEASUREMENTS

I_1, transverse (5), mean 3.9 ± .1, OR 3.7–4.2, CV 5.1. P_4, AP(10), mean 9.4 ± .1, OR 8.6–10.3, CV 5.3; BL(8), mean 5.0 ± .2, OR 4.3–5.7, CV 10.0. M_1, or M_2, AP(6), mean 3.9 ± .1, OR 3.7–4.4, CV 7.7; BL(7), mean 3.3 ± .1, OR 2.8–3.8, CV 9.1. M_3, AP(1), 2.5; BL(1), 2.1. P^4, AP(9), mean 8.5 ± .3, OR 7.3–10.2, CV 11.8; BL(9), mean 5.4 ± .2, OR 4.5–6.2, CV 11.1. M^1 or

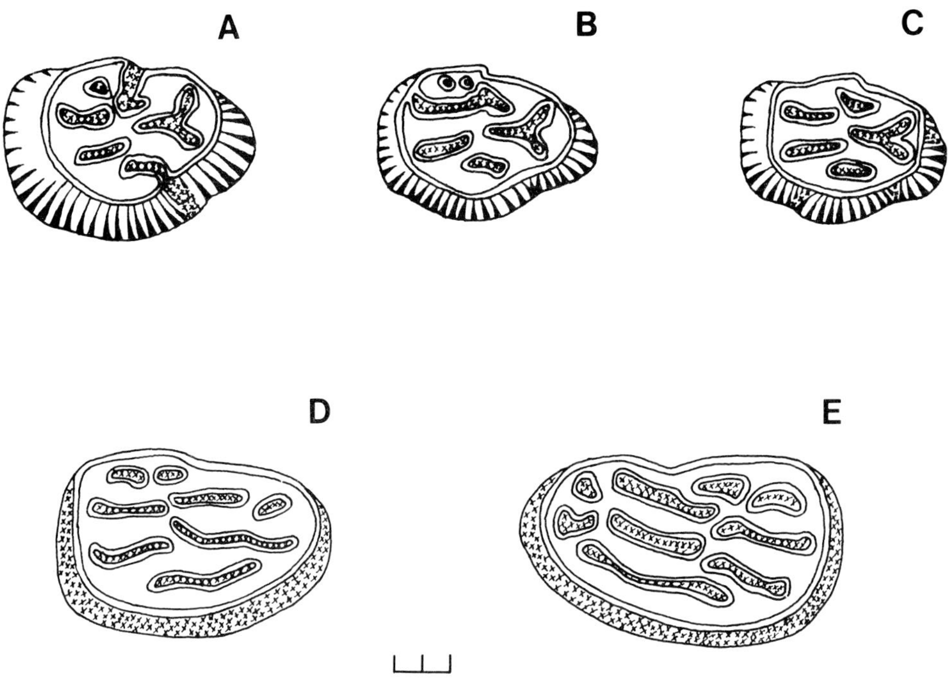

Fig. 50 *Mylagaulus* cf. *M. laevis*, upper premolars in occlusal aspect. Buccal to top of page. Scale in mm.
A. RP^4, SMNH P123, anterior to right.
B. RP^4, ROM 7289, anterior to right.
C. RP^4, SMNH P122, anterior to right.
D. RP^4, NMC 10026, anterior to right.
E. LP^4, NMC 9432, anterior to left.

M^2, AP(2), OR 2.5–3.1; BL(1), 3.4. M^3, AP(3), mean 2.5, OR 2.4–2.7; BL(3), mean 2.5, OR 2.5–2.6.

DISCUSSION

Terminology (Fig. 47) follows Black and Wood (1956), except for valleys on the lingual sides of lower premolars; the feature they call the metaflexid I refer to as entoflexid, because of its association with the entoconid, and I call the anterolingual valley metaflexid, because it is close to the metaconid.

The holotype skull and mandible (AMNH 9043) of *Mylagaulus laevis*, from the Pawnee Creek Beds, has P^4s which match those of ROM 7292, SMNH P122, and NMC 10026, and P_4 which matches all Wood Mountain P_4s. A referred skull and mandible (AMNH 17576) from the Lower Snake Creek Beds, matches in P^4 ROM 7289, SMNH P122, and P391, and in P_4 all the Wood Mountain specimens.

P^4 of the Wood Mountain mylagaulid shows great variability in size, and values of CV above are high for P^4. The entire Wood Mountain size range is expressed in type or referred specimens of *Mylagaulus laevis*, however. It is clear that both *M. laevis* and the Wood Mountain form show far greater variability in P^4 than in P_4, and it may be that a size-bimodality exists in a large sample of P^4.

NMC 8968, a mandibular fragment with P_4, matches the holotype of *M. laevis* in the position of the mental foramen, position and strength of the masseteric ridge, and size and pattern of P_4 (Fig. 48E).

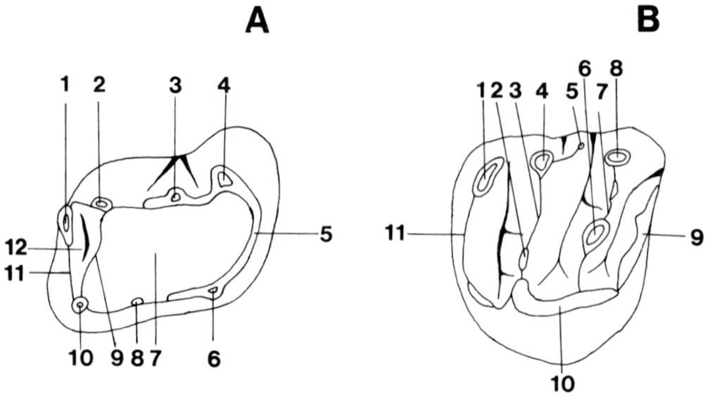

Fig. 51 Sciuridae. Terminology used in description of cheek teeth. Anterior to left, buccal to top of page. Not to scale.

A. RP_4: 1, anteroconid; 2, protoconid; 3, mesoconid on ectolophid; 4, hypoconid; 5, posterolophid; 6, entoconid; 7, talonid basin; 8, mesostylid; 9, metalophid; 10, metaconid; 11, anterior cingulum; 12, trigonid basin.

B. LM^1 or M^2: 1, parastyle; 2, protoconule; 3, protoloph; 4, paracone; 5, mesostyle; 6, metaconule; 7, metaloph; 8, metacone; 9, posterior cingulum; 10, protocone; 11, anterior cingulum.

Family Sciuridae

Spermophilus sp.

Figs. 52; 55A, B

REFERRED SPECIMENS. RP_4 (ROM 7192–7194), LM_1 (ROM 7199, 7202, 7203), RM_1 (ROM 7205), LM_2 (ROM 7195–7198, 7200, 7201), RM_2 (ROM 7206, 7207), LM_1 or M_2 (ROM 7204), LM_3 (ROM 7208, 7251), LP^4 (ROM 7209), LM^1 or M^2 (ROM 7210–7215), RM^1 or M^2 (ROM 7216–7221), LM^3 (ROM 7222, 7223), RM^3 (ROM 7224).

DESCRIPTION

P_4 (Fig. 52A) with elongate talonid basin taking up most of crown; trigonid basin completely enclosed, and metalophid complete; proportionally greater expansion than in P_4 of *Tamias* sp. (Fig. 53A).

M_1 (Fig. 52B) showing some reduction of posterolingual corner; trigonid valley running anterobuccad to anterior surface of crown; metalophid of ROM 7202 complete; shape of M_1 varying as that of M_2.

M_2, as in most species of *Spermophilus* figured by Black (1963b), distinguished from M_1 by larger size; in ROM 7200 (Figs. 52C, 55A), no reduction of posterolingual corner of crown seen; in ROM 7195 (Fig. 52D), posterolingual corner reduced; in ROM 7198 (Fig. 52E), greatly reduced.

ROM 7200, M_2 (Figs. 52C, 55A), showing little anteroposterior compression, tooth being somewhat wider than long; other figured specimens, ROM 7195 and 7198 (Fig. 52D, E), significantly wider than long; in ROM 7200 and 7195 (Fig. 52C, D), metalophid complete, trigonid basin not joining talonid basin; in ROM 7195, however, trigonid basin surrounded by cingula, whereas in ROM 7200 trigonid basin weakly connected with anterior crown surface; in ROM 7198 (Fig. 52E), metalophid incomplete, trigonid basin joining talonid basin, almost reaching anterior surface.

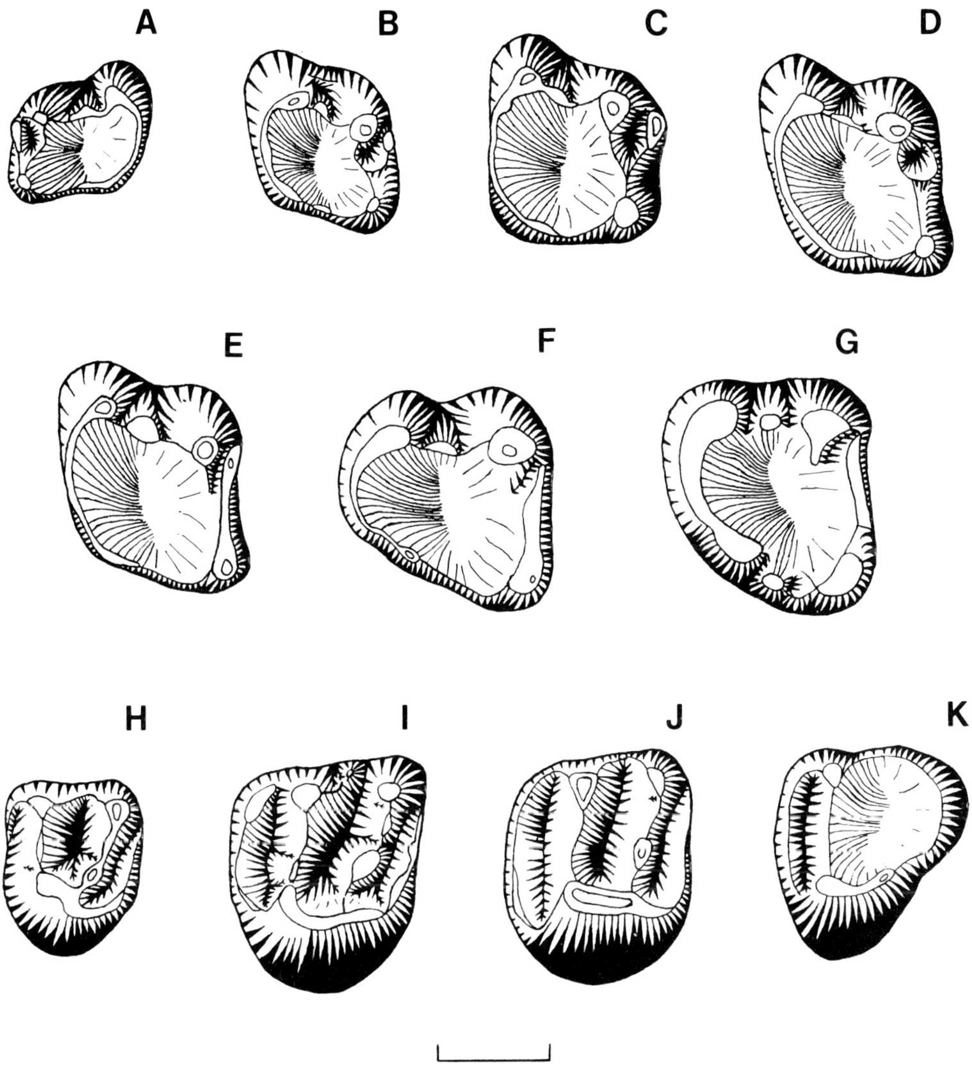

Fig. 52 *Spermophilus* sp., cheek teeth in occlusal aspect. Buccal to top of page. Scale 1 mm.

A. RP$_4$, ROM 7194, anterior to left.
B. LM$_1$, ROM 7202, anterior to right.
C. LM$_2$, ROM 7200, anterior to right.
D. LM$_2$, ROM 7195, anterior to right.
E. LM$_2$, ROM 7198, anterior to right.
F. LM$_3$, ROM 7208, anterior to right.
G. LM$_3$, ROM 7251, anterior to right.
H. LP^4, ROM 7209, anterior to left.
I. LM^1 or M^2, ROM 7211, anterior to left.
J. LM^1 or M^2, ROM 7212, anterior to left.
K. LM^3, ROM 7222, anterior to left.

Two figured specimens of M_3 differing in several respects; ROM 7208 (Fig. 52F) about as wide as long, with no visible mesostylid, resembling in proportion M_3 of *Spermophilus matthewi* (Black, 1963b, pl. 18, fig. 2B); ROM 7251 (Fig. 52G), however, much wider than long, with prominent mesostylid, showing greater reduction of posterolingual corner of crown than ROM 7208; in both specimens, trigonid basin meets talonid basin, nearly reaching anterior surface of crown.

In summary, lower molars of *Spermophilus* sp. showing reduced posterolingual corners, generally showing marked anteroposterior compression; lower cheek teeth with very small (Fig. 52B) to large and prominent (Fig. 52G) mesostylids; in some specimens mesostylid absent (Fig. 52E, F); in most, mesoconid, if present, early attached to cingulum formed by entoconid (Fig. 52A, C, D); entoconid not distinct from posterolophid in worn specimens.

Upper cheek teeth (Figs. 52H, I, J, K; 55B) with complete or nearly complete protolophs; large, low metaconules of P^4 through M^2 near bases of protocones; parastyles poorly developed or absent on most specimens.

P^4, ROM 7209 (Fig. 52H), showing only short ridge in position of parastyle; short valley running obliquely down anterior face of crown; long, thin, laterally directed protoconule close to base of protocone; protoloph complete; no mesostyle; metaloph complete; posterior cingulum enclosing laterally oriented valley, meeting posterior surface of crown posterior to metacone.

ROM 7211 (Figs. 52I, 55B) with well-developed parastyle, but other figured M^1 or M^2, ROM 7212 (Fig. 52J), not showing parastyle; small protoconule in ROM 7211, but not in ROM 7212, both specimens with distinct mesostyles; metalophs of both specimens complete; ROM 7211 with cuspule just lingual to metacone, not shown in ROM 7212.

M^3 (Fig. 52K) with no protoconule, only paracone and protocone; large posterior basin not broken by ridge, cuspule, or valley.

MEASUREMENTS

P_4, AP(3), mean 1.39, OR 1.32–1.47; BL(3), mean 1.32, OR 1.28–1.36. M_1, AP(4), mean 1.32, OR 1.24–1.41; BL(4), mean 1.55, OR 1.52–1.58. M_2, AP(7), mean 1.74 ± .04, OR 1.60–1.90, CV 5.7; BL(7), mean 1.96 ± .03, OR 1.85–2.09, CV 4.1. M_3, AP(2), OR 1.93–1.96; BL(2), OR 1.94–2.07. P^4, AP(1), 1.25; BL(1), 1.54. M^1 or M^2, AP(10), mean 1.63 ± .04, OR 1.32–1.80, CV 8.0; BL(11), mean 2.02 ± .04, OR 1.77–2.24, CV 7.4. M^3, AP(2), OR 1.48–1.79; BL(3), mean 1.67, OR 1.51–1.85.

DISCUSSION

I have followed the ruling of the Committee on Nomenclature, American Society of Mammalogists (1968) in the use of the generic name *Spermophilus* rather than *Citellus*. According to the Committee, the use of *Spermophilus* conforms with the International Rules for Zoological Nomenclature.

The condition of the upper molars from the Wood Mountain Formation, with low protolophs and long metalophs in M^1 and M^2, indicates that the Wood Mountain *Spermophilus* is in the *Otospermophilus* stage of development (Black, 1963b, p. 195). Black (pers. comm., 1968) states that the lower cheek teeth are more advanced in the reduction of the posterolingual corners and in the degree of anteroposterior compression than previously described Miocene *Spermophilus*. Both of these trends are better developed than in *S. primitivus* (Black, 1963b, pl. 18, fig. 1) and are developed to about the same degree as in *S. matthewi* of the Clarendonian (Black, 1963b, pl. 18, fig. 2).

Tamias sp.
Figs. 53; 55C, D

REFERRED SPECIMENS. LP_4 (ROM 7227), LM_1 or M_2 (ROM 7228, 7229), RM_1 or M_2 (ROM 7230–7234), RM_3 (ROM 7235), LP^4 (ROM 7236), RP^4 (ROM 7237, 7238, 7250), LM^1 or M^2 (ROM 7239, 7240), RM^1 or M^2 (ROM 7241–7243).

TENTATIVELY REFERRED. L. mandibular fragment with LI_1, alveolus for LP_4 (ROM 1739).

DESCRIPTION

Lower cheek teeth (Figs. 53A–E; 55C) with angular posterolingual corners and no anteroposterior compression; entoconid forming cingulum, usually clearly separated from posterolophid; minute mesostylid joining entoconid early in wear, showing clearly only in nearly unworn specimens.

P_4 (Fig. 53A) showing metaconid and protoconid close together, separated by anteriorly sloping valley; metaconid broken in figured specimen, but apparently larger, higher of two cusps; slight swelling at centre of ectolophid possibly representing small mesoconid, but probably not.

In M_1 or M_2, posterolingual corner rounded at tip (Fig. 53B, C) or sharply angular (Figs. 53D, 55C); posterolingual corner never reduced and gently curved as in *Spermophilus* sp. (Fig. 52E).

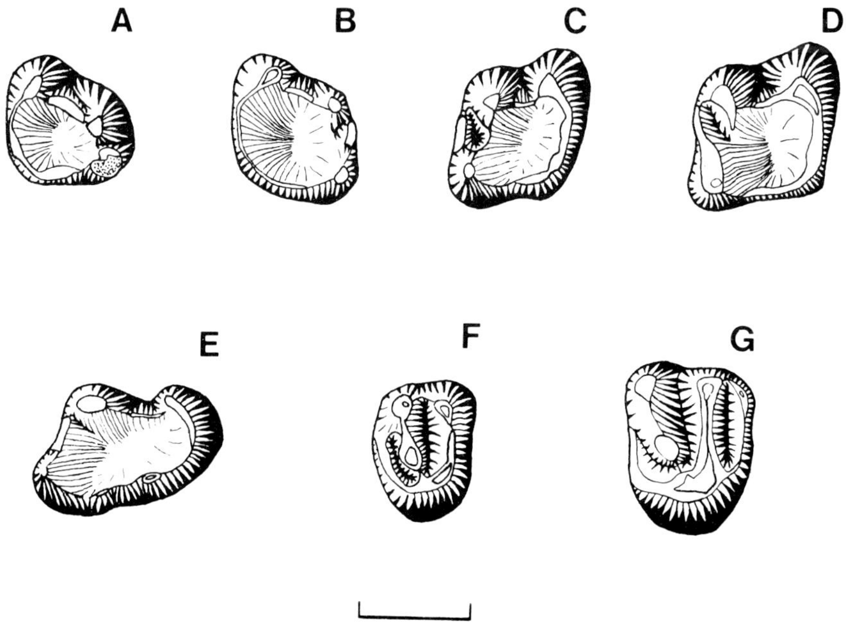

Fig. 53 *Tamias* sp., cheek teeth in occlusal aspect. Buccal to top of page. Scale 1 mm.
A. LP_4, ROM 7227, anterior to right.
B. LM_1 or M_2, ROM 7228, anterior to right.
C. RM_1 or M_2, ROM 7230, anterior to left.
D. RM_1 or M_2, ROM 7234, anterior to left.
E. RM_3, ROM 7235, anterior to left.
F. RP^4, ROM 7238, anterior to right.
G. RM^1 or M^2, ROM 7242, anterior to right.

Metalophids of M_1 or M_2 complete (Fig. 53C) or incomplete (Figs. 53B, D; 55C); trigonid basin sloping posterad into talonid basin as in ROM 7230 (Fig. 53E), or connecting weakly with anterior edge of tooth as in ROM 7228 (Fig. 53B); M_1 and M_2 not distinct in size and shape.

M_3 narrow, elongate; posterolophid curving far buccad beyond line of metalophid (Fig. 53E); entoconid represented by small cusp; mesostylid visible in almost unworn tooth.

Lower jaw fragment (not shown) referred to *Tamias* because lower incisor striated as in many sciurids, size smaller than would be expected for *Spermophilus*, diastema not dipping sharply in front of P_4; according to Black (1963b, p. 128), in members of tribe Tamiini, dorsal border of ramus relatively level anterior to P_4, not dipping sharply.

In upper cheek teeth (Figs. 53F, G; 55D), metaconules large; metalophs constricted at connections to protocones; anterior cingula running from bases of protocones to anterobuccal corners of teeth; no parastyle in P^4 (Fig. 53F); many upper molars with weak parastyles (Figs. 53G; 55D); mesostyles large but low (Figs. 53F, G; 55D).

MEASUREMENTS

P_4, AP(1), 1.14; BL(1), 1.13. M_1 or M_2, AP(7), mean 1.24 ± .03, OR 1.09–1.35, CV 6.5; BL(7), mean 1.40 ± .02, OR 1.31–1.48, CV 4.3. M_3, AP(1), 1.56; BL(1), 1.17. P^4, AP(4), mean 0.97, OR 0.92–1.03; BL(4), mean 1.26, OR 1.24–1.28. M^1 or M^2, AP(5), mean 1.27 ± .04, OR 1.19–1.45, CV 7.1; BL(5), mean 1.66 ± .05, OR 1.50–1.83, CV 6.6.

DISCUSSION

In size and pattern of teeth, *Tamias* sp. from Wood Mountain resembles *T.* sp. from Martin Canyon Quarry A (Black, 1963*b*, pl. 1, fig. 2) and other Arikareean and Hemingfordian *Tamias* in showing no reduction of the posterolingual corner of the crown in lower molars. ROM 7227 (Fig. 53A) is similar in size, shape, and cusp arrangement to the specimen figured by Klingener (1968, fig. 3F) as *Tamias* sp. from the Norden Bridge local fauna. The Norden Bridge material is not sufficient to establish whether the taxon is conspecific with the Wood Mountain *Tamias*.

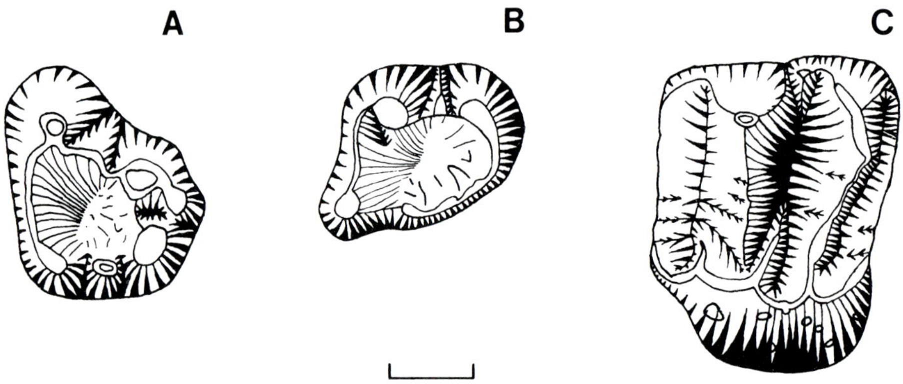

Fig. 54 *Sciuropterus* sp., cheek teeth in occlusal aspect. Buccal to top of page. Scale 1 mm.

A. LP_4, ROM 7244, anterior to right.

B. RM_3, ROM 7246, anterior to left.

C. LM^1 or M^2, ROM 7249, anterior to left.

Sciuropterus sp.

Figs. 54; 55E, F

REFERRED SPECIMENS. LP_4 (ROM 7244), LM_1 or M_2 (ROM 7245), RM_3 (ROM 7246, 7247), LP^4 (ROM 7248, 7252), LM^1 or M^2 (ROM 7249).

DESCRIPTION

ROM 7244, lower premolar (Figs. 54A, 55E), marked by large distinct cusps and deep valleys to greater degree than third lower molar (Fig. 54B); large mesoconid clearly demarcated by deep valleys; mesostylid, far larger than in other Wood Mountain sciurids, separated from metaconid and entoconid by deep valleys; metalophid complete, separating trigonid basin from talonid basin.

P_4 (Figs. 54A, 55E) showing well-developed anteroconid, connected directly to protoconid, and connected to metaconid by strong anterior cingulum, separating trigonid basin from anterior surface of crown.

ROM 7246 (Fig. 54B), third lower molar, showing large metaconid; mesostylid small, joined to entoconid; metalophid incomplete, trigonid basin joining talonid basin.

In lower cheek teeth, complex crenulations in talonid basins, better shown in Fig. 55E than Fig. 54A and B ; crenulations not as well developed as in *Sciuropterus mathewsi* (James, 1963, fig. 37).

ROM 7249 (Figs. 54C, 55F), M^1 or M^2, resembling corresponding teeth of *Sciuropterus uphami* (James, 1963, fig. 35); protocone elongate, protoloph and metaloph complete; paraconule indistinct; protoloph constricted at connection to protocone; posterior cingulum bearing low, cingular hypocone; anterior and posterior cingula enclosing shallow, crenulate basins; parastyle not apparent; accessory lophs running from protocone into both central and anterior valleys; mesostyle low. Height of crown of ROM 7249 much like that of *S. uphami* (James, 1963, fig. 35), much less than that of *S. mathewsi* from Cuyama Basin (James, 1963, fig. 36).

Teeth of *Sciuropterus* from Wood Mountain Formation larger than corresponding teeth of *S. uphami*, *S. mathewsi* (James, 1963, tables 15, 16), far smaller than teeth of *Sciuropterus* sp. from Red Basin, Oregon (Shotwell, 1968, p. 25); size similar to teeth of *S. jamesi* (Lindsay, 1972, table 6).

MEASUREMENTS

P_4, AP(1), 2.46; BL(1), 2.68. M_1 or M_2, BL(1), 2.73. M_3, AP(1), 2.46; BL(1), 2.11. P^4, AP(1), 2.75; BL(2), OR 2.62–2.86. M^1 or M^2, AP(1), 2.90; BL(1), 3.74.

DISCUSSION

These specimens represent the first record of a flying squirrel from the Tertiary of the Great Plains. The Wood Mountain material is not sufficient for the erection of a new species. It is interesting, however, that teeth of the Wood Mountain form show features of pattern and height of crown similar to those of Upper Miocene species from the Cuyama Basin of California.

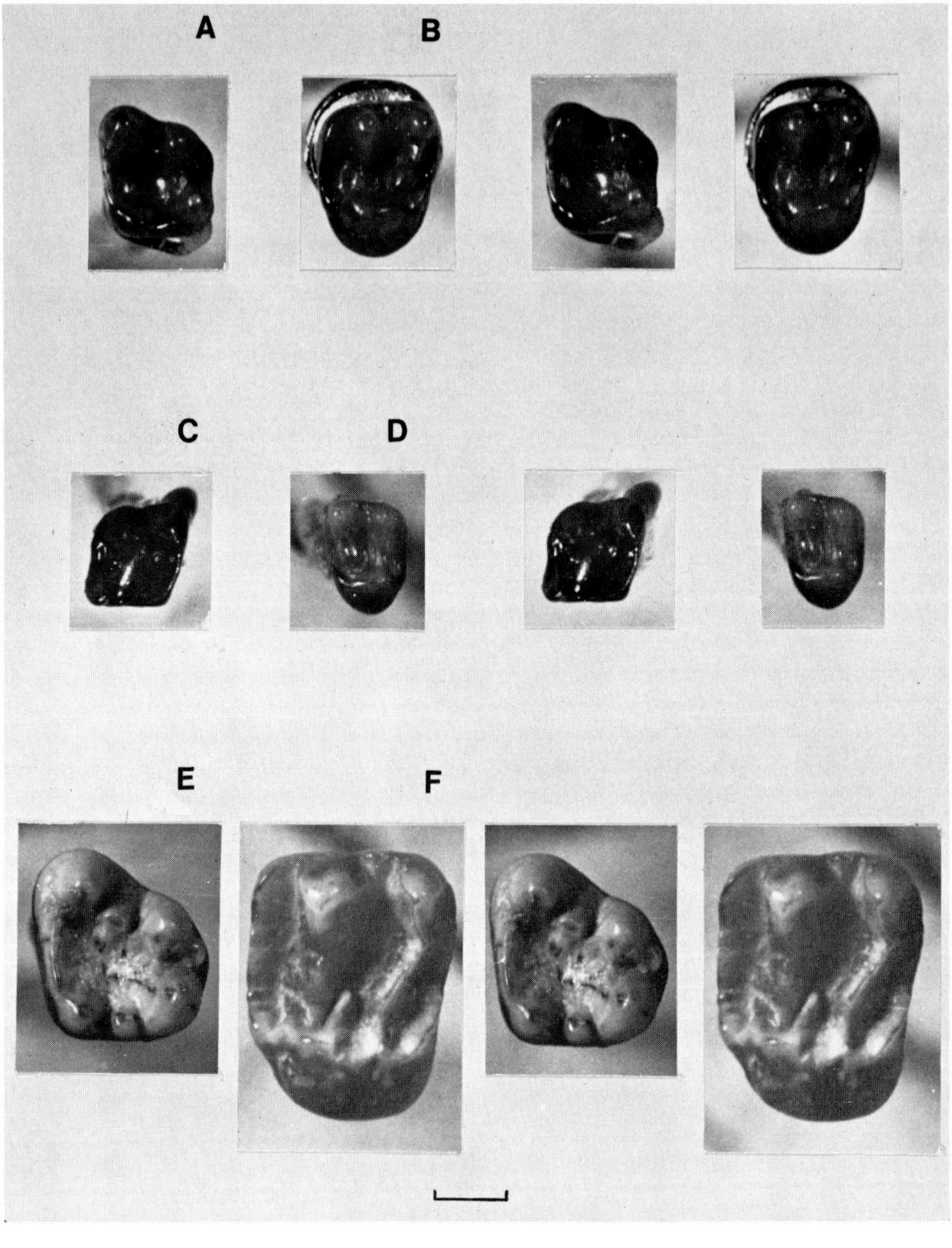

Fig. 55 *Spermophilus, Tamias,* and *Sciuropterus,* cheek teeth in occlusal aspect. Buccal to top of page. Scale 1 mm.

A,B. *Spermophilus* sp.

A. LM_2, ROM 7200, anterior to right.

B. LM^1 or M^2, ROM 7211, anterior to left.

C,D. *Tamias* sp.

C. RM_1 or M_2, ROM 7234, anterior to left.

D. RM^1 or M^2, ROM 7242, anterior to right.

E,F. *Sciuropterus* sp.

E. LP_4, ROM 7244, anterior to right.

F. LM^1 or M^2, ROM 7249, anterior to left.

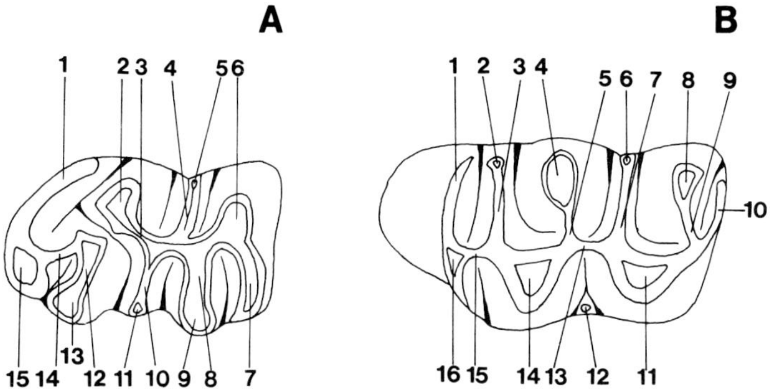

Fig. 56. *Copemys kelloggae canadensis*. Terminology used in description of cheek teeth. Anterior to left, buccal to top of page. Not to scale.

A. RM_1: 1, anterior cingulum; 2, protoconid; 3, central mure; 4, ectolophid; 5, ectostylid; 6, hypoconid; 7, posterior cingulum; 8, hypolophid; 9, entoconid; 10, mesolophid; 11, mesostylid; 12, metalophid; 13, metaconid; 14, anterior mure; 15, anteroconid.

B. LM^1: 1, anterior cingulum; 2, parastyle; 3, anterior protoloph; 4, paracone; 5, posterior protoloph; 6, mesostyle; 7, mesoloph; 8, metacone; 9, metaloph; 10, posterior cingulum; 11, hypocone; 12, enterostyle; 13, central mure; 14, protocone; 15, anterior mure; 16, anterocone.

Family Cricetidae

Copemys kelloggae canadensis subsp. nov.

Figs. 56–58

HOLOTYPE. L. mandibular fragment with LM_{1-2} (ROM 6589).

PARATYPES. LM_1 (ROM 6502 6522, 6638 6641), RM_1 (ROM 6524, 6525, 6527–6530, 6532–6535, 6537–6540, 6542, 6543, 6546, 6642, 6643).

REFERRED SPECIMENS. Incisors (ROM 6636, 6637), LM_1 (ROM 6501, 6523), L. mandibular fragment with LM_1 (ROM 6590), RM_1 (ROM 6526, 6536, 6541, 6544, 6545), L. mandibular fragment (ROM 6584–6588), R. mandibular fragment (ROM 6651), LM_2 (ROM 6547–6561), RM_2 (ROM 6562–6576). LM_3 (ROM 6531, 6577, 6578), RM_3 (ROM 6579–6583), LM^1 (ROM 6591–6604, 6644–6647), RM^1 (ROM 6605–6621, 6648–6650), LM^2 (ROM 6622–6628), RM^2 (ROM 6629, 6630, 6634), LM^3 (ROM 6631–6633), RM^3 (ROM 6635).

TYPE LOCALITY. Kleinfelder Farm Locality, Lsd. 16, Sect. 31, Twp. 2, Rge 30, W. 2nd Merid., near Rockglen, Saskatchewan.

DIAGNOSIS OF SUBSPECIES. Size, height of crown, pattern of teeth comparable to that in *C. kelloggae kelloggae* (Hoffmeister), 1959, but frequency of ectolophid in M_1 about 50%.

DESCRIPTION

Of the holotype and paratypes, 45 first lower molars clearly showing presence or absence of ectolophid, 25 with ectolophid (in 2, ectolophid not reaching ectostylid), and 20 with no ectolophid. In sample of nominal subspecies, *C. kelloggae kelloggae* from Nebraska, only one of nine M_1s with ectolophid

(Klingener, 1968, p. 71). On basis of this difference, Wood Mountain population designated as separate subspecies.

Upper and lower cheek teeth of *C. k. canadensis* varying greatly in size (see measurements) and height of crown (Fig. 57A, B). Considerable variability shown in degree of development of accessory cusps, lophs; Table 12, after Hooper (1957) and Shotwell (1967a), with data necessary for comparison of *C. k. canadensis* with other fossil *Copemys*. In *C. k. canadensis* accessory cusps and lophs more frequently developed than in any other known fossil *Copemys* species; especially high frequencies of development of ectostylid-ectolophid complex in lower molars and parastyle-anterior protoloph complex in M^1.

M_1 ranging from short, proportionally narrow teeth (Fig. 58A), to large narrow teeth (Fig. 58B), to proportionally broader teeth (Fig. 58C); degree of alternation of principal cusps varying little; connection of entoconid to central mure often via hypolophid directed buccad, reaching central mure just anterior to hypoconid (Fig. 58C, D); same mode of connection shown in ROM 6534 (Fig. 58E); in ROM 6514 (Fig. 58A), hypolophid making sharp turn, buccal portion directed anterobuccad, meeting central mure midway between hypoconid and mesolophid; in ROM 6641 (Fig. 58B), hypolophid straight, and directed anterobuccad, meeting central mure midway between hypoconid and mesolophid.

Metalophid connected to protoconid in comparatively little-worn specimens (Fig. 58A, C), or separated from protoconid by shallow valley in well-worn specimens (Fig. 58B, D, E). Metalophid connected to anterior portion of tooth directly through anteroconid (Fig. 58B, C, E), or with anterior cingulum just lingual to anteroconid (Fig. 58A, D; Klingener, 1968, fig. 3A); in some specimens, connection very weak (Klingener, 1968, fig. 3B, C).

Laterally directed valleys of M_1 connecting directly to lateral surface of crown, or blocked by styles or ridges; each valley blocked at least once in illustrations (Fig. 58A–E); styles sometimes prominent but not blocking valleys completely, as large mesostylid in ROM 6534 (Fig. 58E).

In M_2 (Fig. 58F), ectolophid occurring in 11 of 28 suitably preserved specimens; proportional width of crown varying as in M_1; cusp connections varying almost as in M_1, but shallow valley never completely separating metalophid and protoconid.

In M_3 (Fig. 58G), 3 of 8 teeth with ectolophids; variation in proportional width as in M_1 and M_2.

Lower jaw (Fig. 57C, D), showing single mental foramen immediately anteroventral to anterior root of M_1; incisor strongly procumbent, not forming sharp angle with horizontal ramus; masseteric crest strong; ascending ramus arising beneath anterior root of M_2.

M^1 strikingly variable in degree of alternation of major cusps; ROM 6649 (Fig. 58H) clearly showing greater alternation of cusps than ROM 6606, 6645, 6596, or 6598 (Fig. 58I, J, K, L); all of above showing greater alternation than ROM 6647 (Fig. 58M).

In most specimens of M^1, mesoloph strong; in ROM 6649 and 6645 (Fig. 58H, J), mesoloph absent or poorly developed; mesostyle generally present; enterostyle present in nearly all specimens, double in ROM 6649 (Fig. 58H); mesostyle absent in ROM 6598 and 6647 (Fig. 58L, M).

Metacone usually sending arm laterad to meet posterior cingulum (Fig. 58I, K, L); in some specimens, arm extending almost directly posterad meeting posterior cingulum near posterobuccal end (Fig. 58H); in one M^1 (Fig. 58M), arm of metacone meeting posterior cingulum just posterobuccal to hypocone, but posterior cingulum ending there, its posterobuccal portion represented by small,

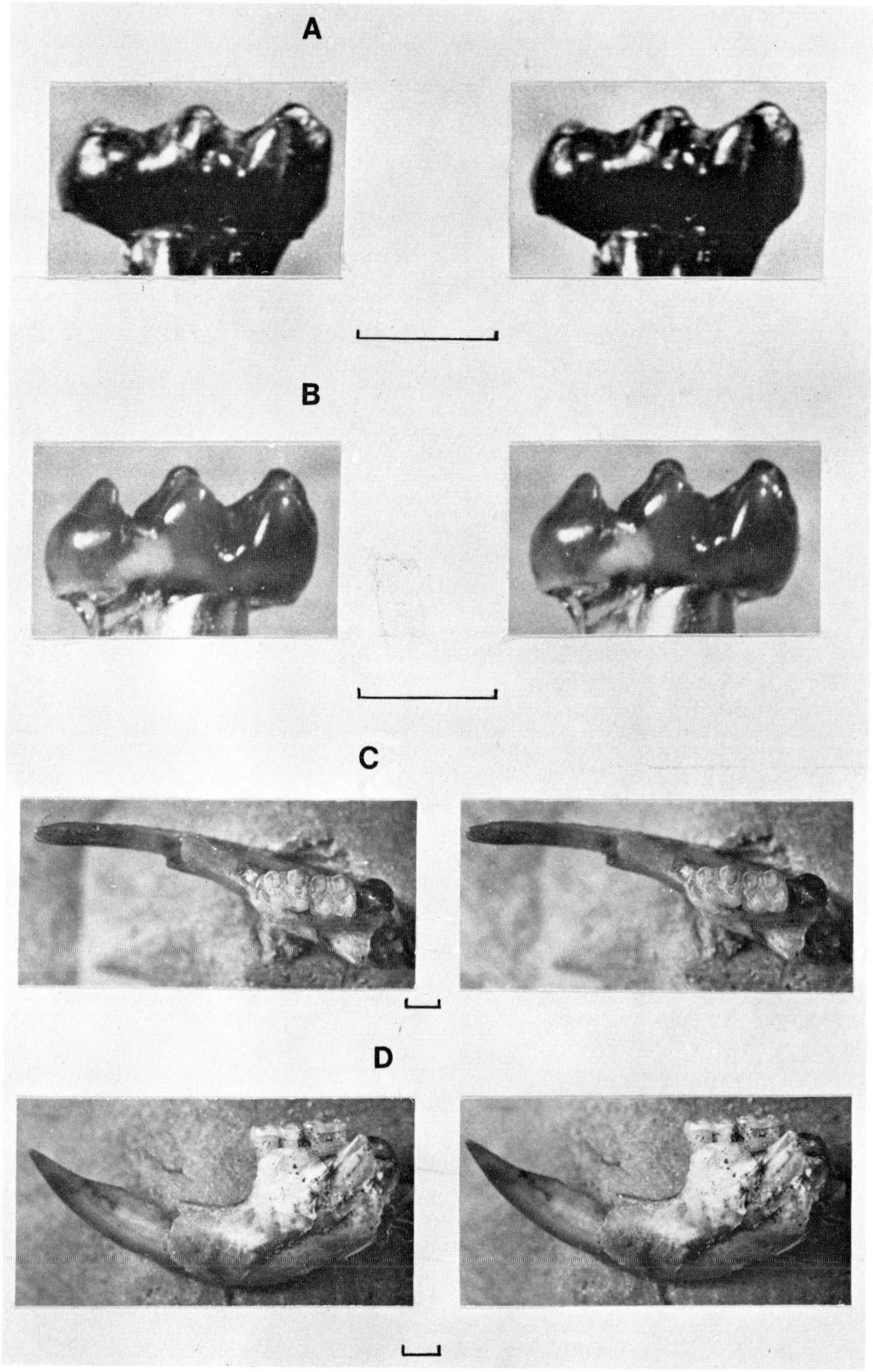

Fig. 57 *Copemys kelloggae canadensis*, molars and mandible. Anterior to left. Scales 1 mm.

A,B. RM^1, lateral aspect, occlusal to top of page.

A. ROM 6606.

B. ROM 6648.

C,D. L. mandible with LM_{1-2}, ROM 6589 (holotype).

C. Occlusal aspect, buccal to bottom of page.

D. Lateral aspect, occlusal to top of page.

low cuspule. Anterocone usually single, occasionally double as in ROM 6596 (Fig. 58K).

M^1 parastyle usually connected to anterior mure by anterior protoloph, present in 24 of 34 well-preserved specimens; in 20 specimens, anterior protoloph connecting with parastyle (Fig. 58K, L); in 2 anterior protoloph prominent but parastyle absent (Fig. 58I); connection of paracone to anterior mure via anterior protoloph, as mentioned by James (1963, p. 120) for *C. russelli*, not seen, but some M^1s showing incipient branch of anterior protoloph directed toward paracone (Fig. 58K, L).

M^2 and M^3 resembling teeth of other *Copemys* species in pattern (Fig. 58N, O); variation in proportional width as in M_1 to M_3 and M^1.

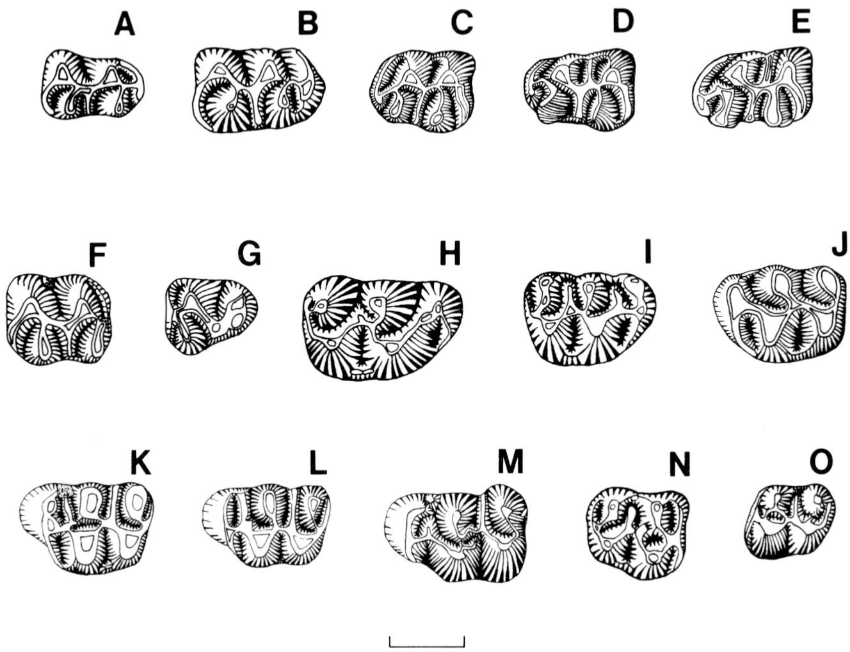

Fig. 58 *Copemys kelloggae canadensis*, cheek teeth in occlusal aspect. Buccal to top of page. Scale 1 mm.

A. LM_1, ROM 6514, anterior to right.
B. LM_1, ROM 6641, anterior to right.
C. RM_1, ROM 6539, anterior to left.
D. RM_1, ROM 6546, anterior to left.
E. RM_1, ROM 6534, anterior to left.
F. LM_2, ROM 6558, anterior to right.
G. RM_3, ROM 6580, anterior to left.
H. RM^1, ROM 6649, anterior to right.
I. RM^1, ROM 6606, anterior to right.
J. LM^1, ROM 6645, anterior to left.
K. LM^1, ROM 6596, anterior to left.
L. LM^1, ROM 6598, anterior to left.
M. LM^1, ROM 6647, anterior to left.
N. RM^2, ROM 6629, anterior to right.
O. LM^3, ROM 6631, anterior to left.

Table 12. Frequency of occurrence of accessory cusps and lophs in *Copemys kelloggae canadensis*.

Accessory cusp and loph	Combination*	M_1	M_2	M_3	M^1	M^2	M^3
Mesostyle-mesoloph or mesostylid-mesolophid	1	.94(45/48)	.97(28/29)		.83(30/36)	.89(8/9)	1.00(4/4)
	2				.03(1/36)	.11(1/9)	
	3		.03(1/29)		.03(1/36)		
	4	.02(1/48)			.06(2/36)		
	5	.04(2/48)			.06(2/36)		
Enterostyle and loph or ectostylid-ectolophid	1	.51(23/45)	.32(9/28)	.38(3/8)			
	2	.04(2/45)	.07(2/28)				
	3						
	4	.22(10/45)	.29(8/28)		.81(30/37)	.78(7/9)	.25(1/4)
	5	.22(10/45)	.32(9/28)	.62(5/8)	.19(7/37)	.22(2/9)	.75(3/4)
Parastyle and anterior protoloph in M^1	1				.59(20/34)		
	2				.06(2/34)		
	3				.06(2/34)		
	4				.18(6/34)		
	5				.12(4/34)		

*1 = Both accessory cusp and loph present and connected to each other.
2 = Both accessory cusp and loph present but not connected to each other.
3 = Only accessory loph present.
4 = Only accessory cusp present.
5 = Neither accessory cusp nor loph present.

MEASUREMENTS

M_1, AP(46), mean 1.55 ± .02, OR 1.36–2.06, CV 9.7; BL met (48), mean 0 98 ± .01, OR 0.75–1.23, CV 9.2; BL hyp (46), mean 1.10 ± .01, OR 0.84–1.28, CV 8.2. M_2, AP(26), mean 1.37 ± .02, OR 121–1.63, CV 6.6; BL met (26), mean 1.13 ± .02, OR 0.92–1.28, CV 7.1; BL hyp (29), mean 1.15 ± .01, OR 0.99–1.30, CV 6.1 M_3, AP(7), mean 1.24 ± .04, OR 1.01–1.36, CV 8.9; BL met (7), mean 1.03 ± .03, OR 0.90–1.14, CV 7.8; BL hyp (8), mean 0.85 ± .03, OR 0.70–0.97, CV 9.4. M^1, AP(34), mean 1.81 ± .03, OR 1.58–2.20, CV 8.3; BL met (37), mean 1.21 ± .02, OR 1.05–1.41, CV 7.4; BL hyp (34), mean 1.21 ± .02, OR 0.97–1.39, CV 9.1. M^2, AP(10), mean 1.35 ± .05, OR 1.11–1.72, CV 11.1; BL met (10), mean 1.22 ± .03, OR 1.10–1.45, CV 8.2; BL hyp (9), mean 1.12 ± .02, OR 1.01–1.23, CV 6.2. M^3, AP(4), mean 1.28, OR 1.17–1.34; BL met (4), mean 1.16, OR 1.06–1.28; BL hyp (4), mean 1.05, OR 0.97–1.10.

DISCUSSION

These specimens are referable to *Copemys* under the diagnosis given by Lindsay (1972, p. 75), because in M_1 the hypolophid does not line up with the central mure, and in M^1 the posterior protoloph does not line up through the central mure with the hypocone. The Wood Mountain population is designated as a new subspecies of *Copemys kelloggae* because of the significantly higher frequency of development of the ectolophid in M_1.

The holotypic lower jaw fragment (Fig. 57C, D) closely resembles two Norden Bridge lower jaws of *Copemys kelloggae kelloggae* (UMMP 53006 and 53032) in the size, shape, and orientation of the incisor, the depth and shape of the ramus, and the position of the mental foramen. The incisor differs in position from that of *C. russelli* from the Cuyama Basin of California, stated by James (1963, p. 127) to be strongly erect.

Comparison of Table 12 with similar tables of Hooper (1957), Shotwell (1967a), and Wolfe and Layne (1968) indicates that *C. k. canadensis* shows as high a degree of complexity as any known living or extinct *Peromyscus* or *Copemys*. James (1963, p. 128), Hooper (1957, p. 48), and Wolfe and Layne (1968, p. 7) suggested that populations with more styles and lophs on cheek teeth may live in denser cover, though no definite selective factors were cited. The great complexity of the dental pattern of *Copemys kelloggae canadensis*, then, may indicate that it inhabited woodland with a dense understory.

Family Zapodidae

Megasminthus tiheni Klingener, 1966

Figs. 60, 61, 62

REFERRED SPECIMENS. Incisors (ROM 6742, 6743), L. mandibular fragment with LI_1 (ROM 6706), LM_1 (ROM 6652–6658, 6667), RM_1 (ROM 6659–6666, 6668–6672), LM_2 (ROM 6673–6679), RM_2 (ROM 6680–6692), LM_3 (ROM 6693–6696, 6698), L. mandibular fragment with LM_3 (ROM 6697), RM_3 (ROM 6699, 6700, 6702), R. mandibular fragment with RM_3 (ROM 6701), L. mandibular fragment with LM_{1-3} (ROM 6703), R. mandibular fragment with RM_{1-2} (ROM 6704), L. mandibular fragment with LM_{2-3} (ROM 6705), partial palate with LP^4, RP^4, LM^1, RM^1 (ROM 6740), RP^4 (ROM 6707), LM^1 (ROM 6708–6711, 6713–6715), L. maxillary fragment with LM^1 (ROM 6712), RM^1 (ROM 6716–6722), LM^2 (ROM 6723–6729), RM^2 (ROM 6730–6737), LM^3 (ROM 6738, 6739), partial palate with alveoli for RP^4, RM^1 (ROM 6741).

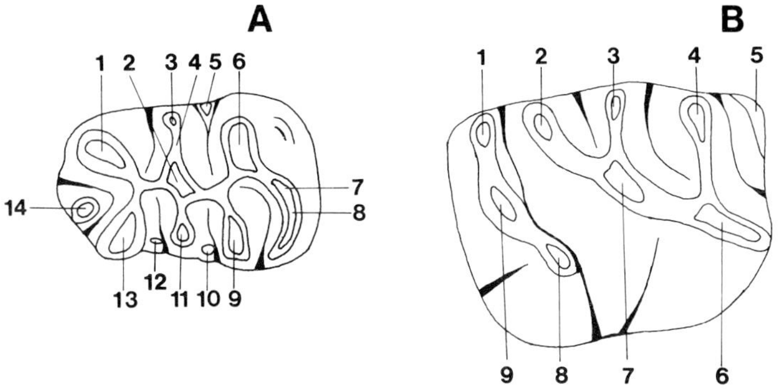

Fig. 59 Zapodidae. Terminology used in description of cheek teeth. Anterior to left, buccal to top of page. Not to scale.

A. RM_1: 1, protoconid; 2, mesoconid; 3, "Gegensporn"; 4, ectolophid; 5, ectostylid; 6, hypoconid; 7, hypoconulid; 8, posterior cingulum; 9, entoconid; 10, entostylid; 11, mesostylid; 12, metastylid; 13, metaconid; 14, anteroconid.

B. LM^1: 1, cusp of anterior cingulum; 2, paracone; 3, mesostyle; 4, metacone; 5, metastyle; 6, hypocone; 7, mesocone; 8, protocone; 9, anterocone.

DESCRIPTION

In occlusal outline of M_1 crown, ROM 6660 and 6663 (Fig. 60A, B) similar to M_1 of tooth row figured by Klingener (1966, fig. 2A); ROM 6667 (Fig. 60C) with marked lateral constriction of metalophid, probably representing extreme variant; M_1 of ROM 6703 (Fig. 61A) intermediate in proportionate width of metalophid.

In both Wood Mountain and Norden Bridge samples, degree of ectolophid development variable; UMMP 52876 (Klingener, 1966, fig. 2A), showing weak connection between "Gegensporn" and ectolophid, ectolophid poorly defined; three of five specimens in UMMP 52875 and 52883 with strong ectolophids, two with low and weak but clearly defined ectolophids; of 19 well-preserved M_1s from Wood Mountain Fauna, 11 with high ectolophids and 8 with low but clearly defined ones; in some specimens ectolophid meeting "Gegensporn" near base, extreme wear necessary to produce dentine lake common to mesoconid and "Gegensporn"; in first lower molars from Norden Bridge and Wood Mountain, about 50% with strong ectolophids.

ROM 6663 with entostylid, metastylid and ectostylid (Fig. 60B). ROM 6663 with low cingulum posterior to hypoconid. Connection of hypoconid to posterior arm of entoconid depending largely on wear; in M_1 of ROM 6703 (Fig. 61A), ROM 6667 (Fig. 60C) connection weaker than in ROM 6660 and 6663 (Fig. 60A, B).

M_2 varying in proportional width: M_2 of ROM 6703 (Fig. 61A) narrower than M_2 of UMMP 52876.

M_3 varying in relative width of hypolophid; M_3 of ROM 6703 with broad hypolophid, M_3 of UMMP 52876 with narrow hypolophid.

Horizontal ramus (Fig. 61B) deep dorsoventrally beneath tooth row, becoming shallow anteriorly; dorsal margin dipping sharply downward anterior to M_1; large, slitlike, dorsally directed mental foramen anteroventral to anterior root of M_1; ascending ramus originating beneath anterior root of M_3; masseteric ridge strong; ramus deepening sharply at angle.

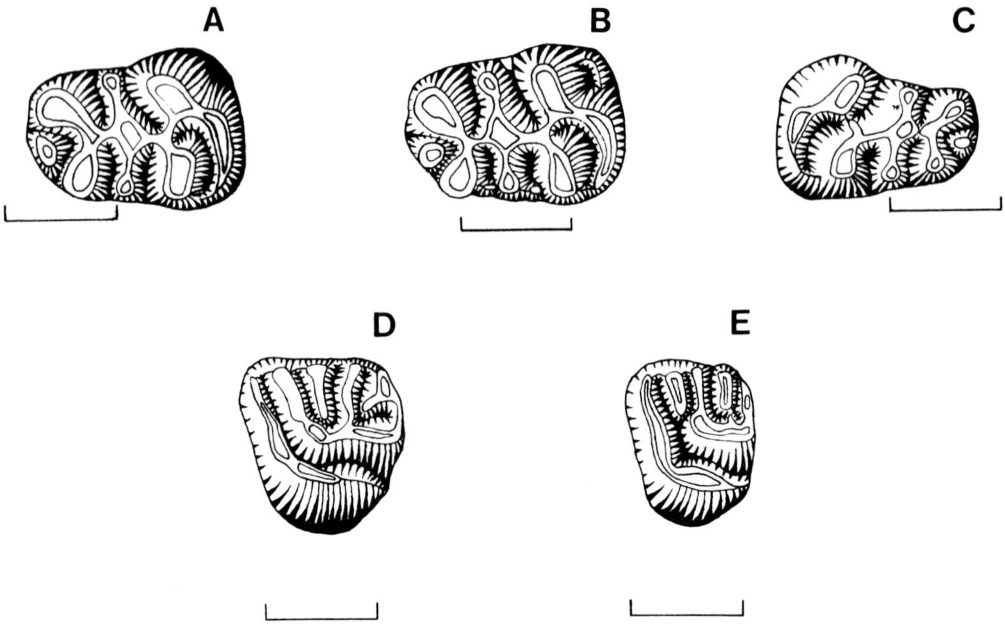

Fig. 60 *Megasminthus tiheni*, cheek teeth in occlusal aspect. Buccal to top of page. Scales 1 mm.

A. RM_1, ROM 6660, anterior to left.
B. RM_1, ROM 6663, anterior to left.
C. LM_1, ROM 6667, anterior to right.
D. LM^3, ROM 6738, anterior to left.
E. LM^3, ROM 6739, anterior to left.

P^4 (Fig. 62) with large central cusp, smaller but distinct lingual and buccal cusps; posteriorly oriented valley running through posterior half of tooth.

M^1 (Fig. 62) and M^2 (not shown) similar to each other and to upper molars described by Klingener (1966, p. 5): "In M^1 the protocone, anterocone, and anterior cingulum are united to form a ridge separated from the hypocone, mesocone, and paracone by a deep valley. This valley is the continuation of the internal re-entrant fold. In ... *Plesiosminthus* this valley is interrupted by a connection between protocone and mesocone. *Zapus* and *Napaeozapus* resemble *Megasminthus* in this character, since in young individuals the valley is usually open. The posterior cingulum in M^1 of *Megasminthus* is almost independent of the hypocone." As in Norden Bridge sample, in some specimens from Wood Mountain weak anterior arm extending from mesocone toward anterior cingulum, but usually posterior wall of anterior cingulum nearly vertical.

Two M^3s, ROM 6738 and 6739 (Fig. 60D, E), differing greatly from each other; both ROM specimens with prominent L-shaped cingula; deep, anteroposteriorly directed valley in ROM 6738 and 6739 separating anterior cingulum and protocone from posterobuccal portion of crown; metacone and paracone separated from each other by mesostyle and accessory loph; metastyle prominent; in ROM 6738 (Fig. 60D), paracone, metacone, and metastyle connected to each other by anteroposteriorly oriented loph, mesostyle separated from anteroposterior loph by shallow valley; in ROM 6739 (Fig. 60E), anteroposterior loph connected to

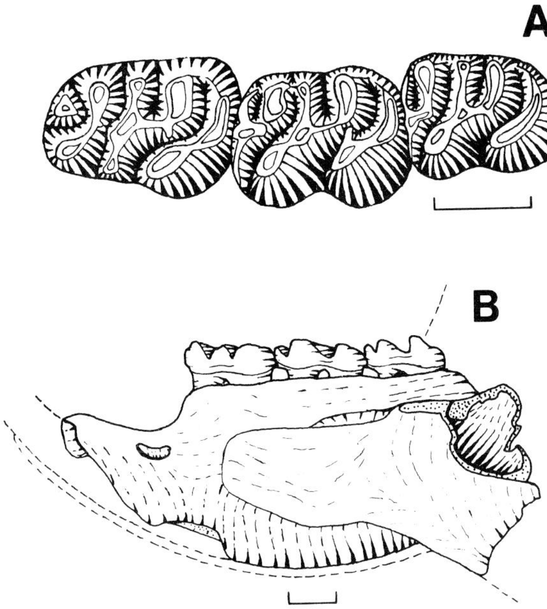

Fig. 61 *Megasminthus tiheni*, L. mandible with LM_{1-3}, ROM 6703. Anterior to left. Scales 1 mm.
A. Occlusal aspect, buccal to bottom of page.
B. Lateral aspect, occlusal to top of page.

mesostyle and metastyle, separated from paracone by shallow valley and from metacone by deep valley.

Palate (Fig. 62) marked anteriorly by two large, connected incisive foramina terminating posteriorly near the centre of P^4; posterior palatine arterial foramina probably even with centre of M^2; inferior zygomatic root originating just anterior to P^4.

MEASUREMENTS

M_1, AP(18), mean 1.93 ± .01, OR 1.78–1.99, CV 3.1; BL met (19), mean 1.18 ± .02, OR 0.97–1.30, CV 7.6; BL hyp (22), mean 1.44 ± .01, OR 1.25–1.53, CV 5.6. M_2, AP(16), mean 1.84 ± .02, OR 1.72–2.00, CV 4.3; BL met (18), mean 1.50 ± .02, OR 1.32–1.66, CV 6.0; BL hyp (23), mean 1.50 ± .02, OR 1.24–1.66, CV 6.7. M_3, AP(7), mean 1.54 ± .05, OR 1.25–1.67, CV 9.1; BL met (17), mean 1.44 ± .04, OR 1.23–1.58, CV 6.9; BL hyp (10), mean 1.22 ± .04, OR 0.97–1.39, CV 9.8. P_4, AP(3), mean 0.79, OR 0.68–0.97; BL(2), OR 1.01–1.01. M^1, AP(14), mean 1.94 ± .01, OR 1.88–1.99, CV 2.1; BL met (15), mean 1.94 ± .02, OR 1.77–2.07, CV 3.6; BL hyp (16), mean 1.84 ± .02, OR 1.56–1.94, CV 4.7. M^2, AP(14), mean 1.64 ± .02, OR 1.50–1.76, CV 4.9; BL met (15), mean 1.76 ± .03, OR 1.57–1.98, CV 6.3; BL hyp (12), mean 1.67 ± .02, OR 1.52–1.85, CV 4.8. M^3, AP(2), OR 1.17–1.41; BL(2), OR 1.47–1.54.

DISCUSSION

This large sample, comprising 90 cheek teeth, is close in most respects to the type and referred material of *M. tiheni* from the Norden Bridge local fauna of

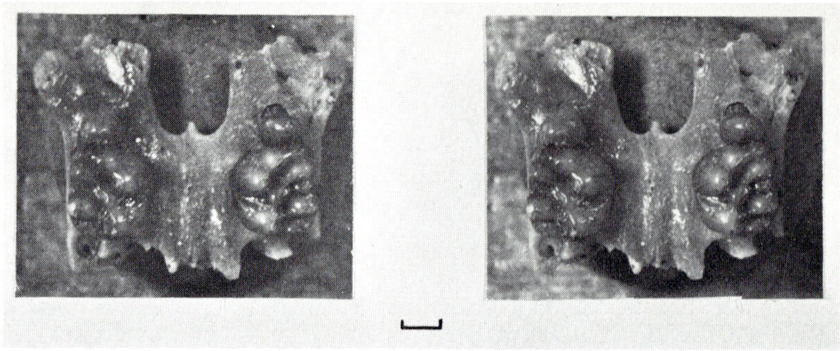

Fig. 62 *Megasminthus tiheni*, partial palate with LP^4, RP^4, LM^1, RM^1, ROM 6740, occlusal aspect. Scale 1 mm.

Nebraska. M_1 from the Wood Mountain sample shows somewhat greater variability than from the Norden Bridge sample. The lower jaw, ROM 6703, does not differ in any discernible way from lower jaws in the University of Michigan collections from the Norden Bridge quarry.

P^4 is identical in the Wood Mountain and Norden Bridge specimens (Fig. 62; Klingener, 1966, fig. 1). The difference which appears in the figures results from different orientations of P^4 in the two palates. P^4 of *Megasminthus tiheni* is entirely different from P^4 in zapodids of the Arikareean and Hemingfordian, which show comparatively larger buccal cusps. The closest approach to *Megasminthus* in the pattern of P^4 is seen in the sicistine *Plesiosminthus galbreathi* (Wilson, 1960, fig. 128).

The only discernible difference between the palates from the Wood Mountain and Norden Bridge Faunas is in the size of the incisive foramina. In the Norden Bridge palate, the incisive foramina "terminate posteriorly at the level of the anterior edge of P^4" (Klingener, 1966, p. 3), while in the Wood Mountain palate they reach posteriorly to the middle of P^4. Aside from this difference, which probably represents individual variation, no other important differences can be seen.

Plesiosminthus sp.

Fig. 63

REFERRED SPECIMENS. RM_1 (ROM 6744–6746), LM_2 (ROM 6747, 6748), RM_2 (ROM 6749), LM^1 or M^2 (ROM 6750), RM^1 or M^2 (ROM 6751).

DESCRIPTION

One M_1, ROM 6744 (Fig. 63A), coinciding almost exactly in size (Table 13) and pattern with M_1 from Norden Bridge local fauna (Klingener, 1966); ROM 6746 (not shown) with the same pattern of connection of major cusps, same double anteroconid, but with large ectolophid.

Three M_2s approximate M_2 figured by Klingener (1966, fig. 2B), but none with double connection of metaconid and protoconid; in ROM 6748 (Fig. 63B), metaconid sending out single, buccally oriented arm, connecting with anterior arm of protoconid just anterolingual to protoconid; other intact M_2, ROM 6749 (not shown), showing connection between anterior arms of protoconid and metaconid taking place just anterobuccal to metaconid, close to anteroconid; arm of metaconid directed anterobuccad.

Upper molars (Fig. 63C) showing pattern reminiscent of that in earlier zapo-

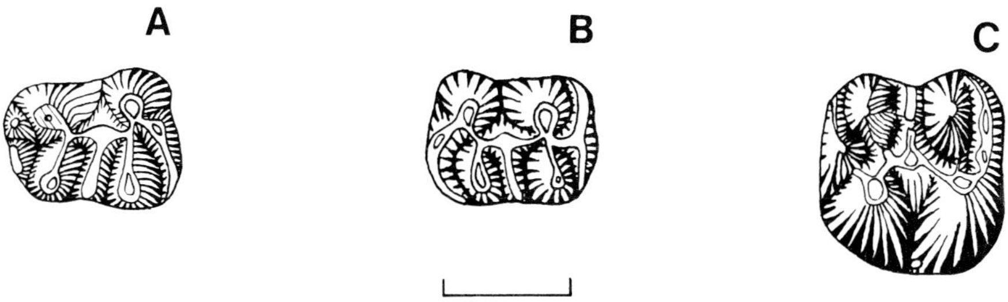

Fig. 63 *Plesiosminthus* sp., cheek teeth in occlusal aspect. Buccal to top of page. Scale 1 mm.
A. RM_1, ROM 6744, anterior to left.
B. LM_2, ROM 6748, anterior to right.
C. LM^1 or M^2, ROM 6750, anterior to left.

dids; figured specimen comparable to upper molars of *Plesiosminthus* (*Schaubeumys*) *sabrae* (Black, 1958, fig. 1A–F), although linearly about 50% larger.

DISCUSSION

Although the sample of *Plesiosminthus* from the Wood Mountain Fauna is small, great variation is seen in the lower molars.

Great similarity is shown to Norden Bridge *Plesiosminthus* sp., both in pattern and in size (Table 13). Neither is known well enough for a species to be named, but with more extensive collections it may be shown that the same species is represented in the Wood Mountain and Norden Bridge Faunas.

Table 13. Measurements of teeth of *Plesiosminthus* sp. from the Wood Mountain Fauna compared with those of specimens from the Norden Bridge Fauna (means for Norden Bridge sample calculated from measurements of Klingener, 1966, p. 2).

Tooth	Dimension	n	Mean	OR	Norden Bridge
M_1	AP	3	1.36	1.28–1.45	1.4 (n = 3)
	BL met	3	0.92	0.89–0.94	
	BL hyp	3	1.01	0.89–1.09	1.1 (n = 3)
M_2	AP	2		1.26–1.32	1.4 (n = 1)
	BL met	2		0.98–1.07	
	BL hyp	3	1.06	1.01–1.10	1.1 (n = 1)
M^1 or M^2	AP	1		1.47	
	BL met	1		1.54	
	BL hyp	2		1.21–1.50	

Family Eomyidae

Pseudadjidaumo russelli (Storer), 1970a

Figs. 65, 66

HOLOTYPE. R. mandibular fragment with RP_4, RM_{1-2} (ROM 7191).

TYPE LOCALITY. Kleinfelder Farm Locality, Lsd. 16, Sect. 31, Twp 2, Rge 30, W. 2nd Merid., near Rockglen, Saskatchewan.

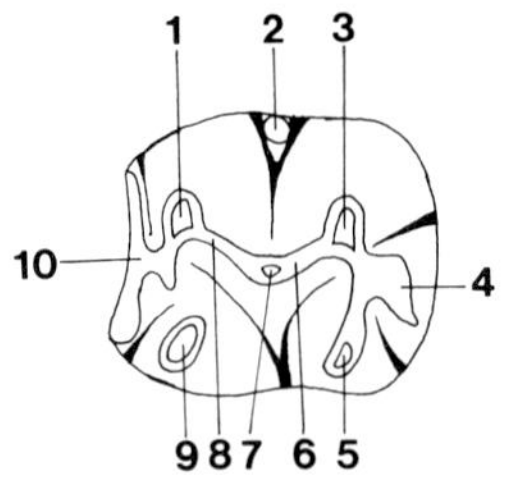

Fig. 64 *Pseudadjidaumo russelli.* RM$_2$, with terminology used in description of cheek teeth. Anterior to left, buccal to top of page. Not to scale.
1, protoconid; 2, ectostylid; 3, hypoconid; 4, posterior cingulum; 5, entoconid; 6, posterior arm of ectolophid; 7, mesoconid; 8, anterior arm of ectolophid; 9, metaconid; 10, anterior cingulum.

DIAGNOSIS. Small eomyid with low-crowned dentition. Lower cheek teeth bunodont. Protoconid and metaconid of P_4 connected by single transverse loph. Mesolophids of lower molars absent.

DESCRIPTION

Lower teeth (Figs. 65; 66A) brachydont; lower premolar with two roots, molars three; tooth patterns composed of four major cusps plus mesoconid; connections between cusps low; mesolophids absent.

Fig. 65 *Pseudadjidaumo russelli,* RP$_4$ to M$_2$, ROM 7191 (holotype), occlusal aspect. Anterior to right, buccal to bottom of page. Scale 1 mm.

Hypoconid and entoconid of P_4 joining, near midline of tooth; posterior cingulum originating near this point of junction, traversing half width of tooth; posterior arm of ectolophid joining hypoconid to rounded mesoconid; anterior arm of ectolophid weak, running anterobuccad toward metaconid; protoconid set slightly anterior to metaconid, with single medial connection.

Metaconid highest cusp in M_1 and M_2, joining protoconid near centre of trigonid; at this junction, anterior cingulum arising, traversing nearly entire width of tooth; rounded mesoconid connected to protoconid and hypoconid by arms of ectolophid; hypoconid and ectolophid joining buccal to midline; small posterior cingulum posterobuccal to hypoconid, traversing less than half width of tooth; small ectostylid in M_2, not seen in M_1.

Lower jaw (Fig. 66B) broken just anterior to P_4, position of mental foramen not shown; posterior edge of mental foramen possibly appearing as part of break surface, in same position as that of *Pseudadjidaumo quartzi* (Shotwell, 1967B, fig. 2A); jaw deepening toward anterior end of tooth row; ascending ramus originating near posterior end of M_2; masseteric fossa, although deep, confined to base of coronoid process.

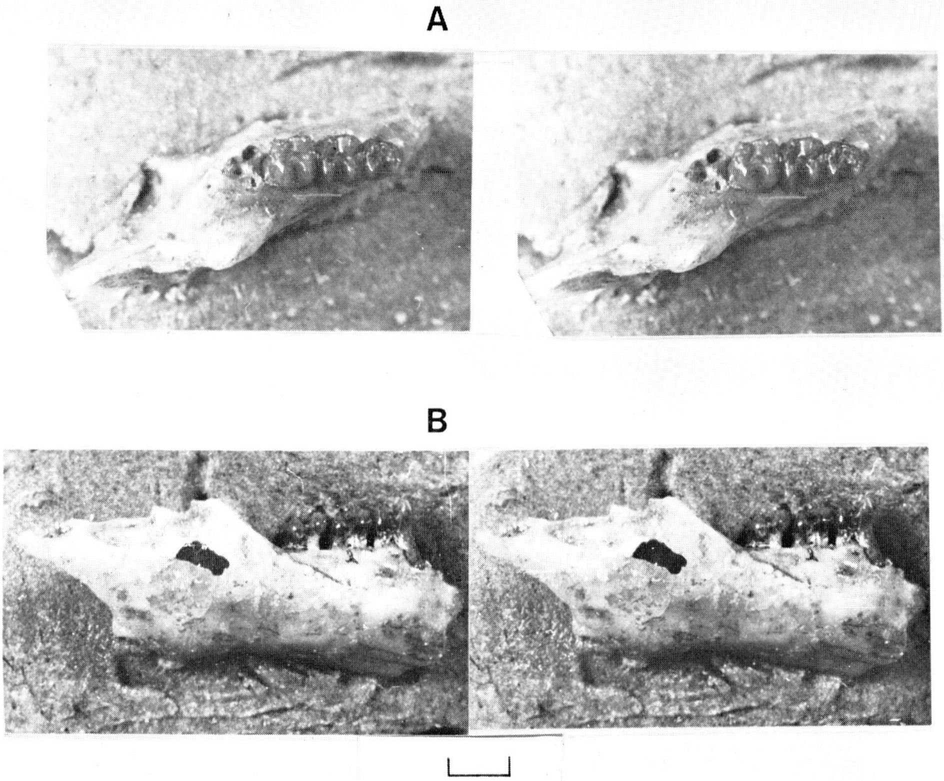

Fig. 66 *Pseudadjidaumo russelli*, R. mandible with RP_4 to M_2, ROM 7191 (holotype). Anterior to right. Scale 1 mm.
A. Occlusal aspect, buccal to bottom of page.
B. Lateral aspect, occlusal to top of page.

MEASUREMENTS

P_4, AP(1), 0.66; BL pro (1), 0.57; BL met (1), 0.65. M_1, AP(1), 0.70; BL met (1); 0.75; BL hyp (1), 0.75. M_2, AP(1), 0.75; BL met (1), 0.76; BL hyp (1), 0.76.

DISCUSSION

The Wood Mountain eomyid is definitely referable to the genus *Pseudadjidaumo* Lindsay, which is said to differ from *Adjidaumo* "in possessing anterior and posterior valleys shallower than the median valley, and the anterior cingulum of lower molars is more reduced labial to the union with the protoconid" (Lindsay, 1972, p. 36). Storer (1970a) included *Pseudadjidaumo russelli* within *Adjidaumo* in expectation of a new genus being erected, because of close resemblance to *Adjidaumo quartzi* (Shotwell, 1967b, fig. 2A), which Lindsay also included in *Pseudadjidaumo*.

P. russelli is clearly separated from other species of *Pseudadjidaumo* by the lack of the mesolophid. The cusps of teeth of *P. russelli* are closer together than those of *P. stirtoni* (Lindsay, 1972, fig. 18) or *P. quartzi* (Shotwell, 1967b, fig. 2A), and the posterior cingulum is narrower buccolingually than in the Norden Bridge specimen (Klingener, 1968, fig. 3E).

Pseudadjidaumo russelli is similar to *Pseudadjidaumo quartzi* from Oregon in size, height of crown, and gross dental morphology. The cusps of teeth of

P. russelli are closer together than those of *P. quartzi*, however, and *P. quartzi* has a poorly-developed mesolophid while *P. russelli* has none.

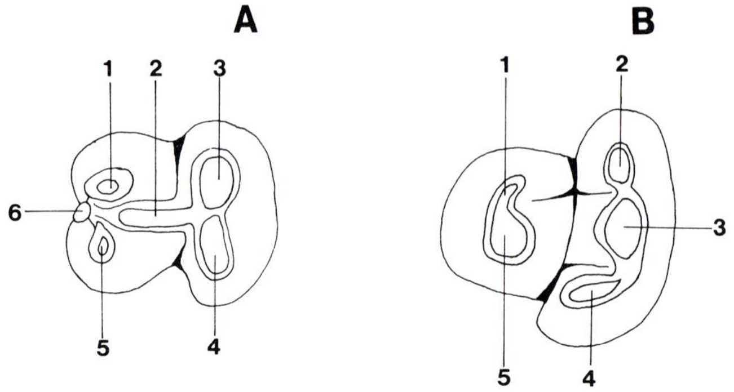

Fig. 67 Heteromyidae. Terminology used in description of premolars. Anterior to left, buccal to top of page. Not to scale.
A. RP_4: 1, protostylid; 2, longitudinal ridge; 3, hypoconid; 4, entoconid; 5, protoconid; 6, anteroconid; small cusp buccal to hypoconid (not shown), hypostylid.
B. LP^4: 1, buccal accessory cusp of protoloph; 2, metacone; 3, hypocone; 4, hypostyle; 5, protocone.

Family Heteromyidae

***Mookomys* sp.**
Figs. 68; 72A

REFERRED SPECIMENS. RP_4 (ROM 7085), LP^4 (ROM 7086, 7087), RP^4 (ROM 7088), LM^1 or M^2 (ROM 7089), RM^1 or M^2 (ROM 7090).

DESCRIPTION

Metalophid of P_4 (Fig. 68A) bearing large entoconid; small cuspule in normal position of hypoconid; protolophid with two cusps, protoconid and protostylid; tooth showing no tendency to form lophs; both transverse and anteroposterior valleys strong; anteroposterior valley blocked by low cingulum at anterior end.

Upper premolar (Figs. 68B, 72A) cuspidate; hypostyle sharply inflected; no cingulum connecting metacone and hypostyle. Upper molars small, cuspidate.

MEASUREMENTS

P_4, AP(1), 0.75; BL pro (1), 0.53; BL met (1), 0.63. P_4, AP(3), mean 1.26, OR 1.23–1.32; BL pro (3), mean 0.61, OR 0.53–0.68; BL met (2), OR 1.01–1.21. M^1 or M^2, AP(2), OR 0.88–0.88; BL pro (2), OR 1.07–1.10; BL met (2), OR 0.96–0.99.

DISCUSSION

Homologies of some cusps in heteromyids are in doubt, especially in premolars. Terminology is a compromise between systems of Rensberger (1971, fig. 67) and Lindsay (1972, fig. 19). Names stress homology of cusps of heteromyids and geomyids (Fig. 74) with each other and with eomyids (Fig. 64) and stress basic similarities of premolars and molars.

Upper premolars are basically similar to corresponding teeth of *Perognathus saskatchewanensis* (Fig. 69D, E). They are smaller, however, comparing in size with P^4 of *Mookomys formicarum*, and the hypostyle is more sharply inflected than in *P. saskatchewanensis*; in *M. formicarum*, the hypostyle does not appear to be inflected (Wood, 1935, fig. 8a).

P_4 shows an unusual feature in being three-cusped, but it is apparent that this condition has come about through the suppression of the hypoconid, and it cannot be ascertained whether the suppression of the hypoconid is a specific or individual feature.

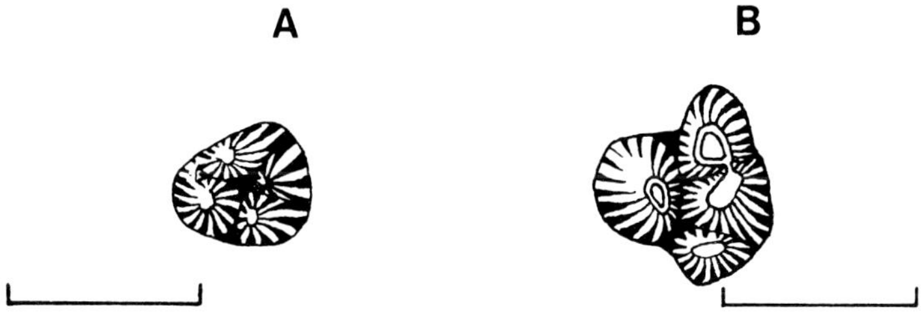

Fig. 68 *Mookomys* sp., premolars in occlusal aspect. Anterior to left, buccal to top of page. Scale 1 mm.
A. RP_4, ROM 7085.
B. LP^4, ROM 7086.

Perognathus saskatchewanensis Storer, 1970a
Figs. 69; 72B, C

HOLOTYPE. LP_4 (ROM 6777).

REFERRED SPECIMENS. LP_4 (ROM 6752–6776, 6778–6786), RP_4 (ROM 6787–6804, 6806–6814, 6816–6818), L. mandibular fragment with broken LP_4 (ROM 6934), LP^4 (ROM 6819–6955), RP^4 (ROM 6856–6872, 6874, 6875, 6877–6902).

TYPE LOCALITY. Kleinfelder Farm Locality, Lsd. 16, Sect. 31, Twp 2, Rge 30, W. 2nd Merid., near Rockglen, Saskatchewan.

DIAGNOSIS. Large *Perognathus*, with P_4 as long as, but slightly narrower than, P_4 of *Perognathus coquorum*; anteroconid present in most specimens of P_4; in most specimens of P^4, ridge extending anterad from middle of metalophid, connecting protolophid and metalophid.

DESCRIPTION

Metalophid of P_4 composed of entoconid and hypoconid; central accessory cuspule (Figs. 69A–C; 72B, C) in conjunction with longitudinal ridge; hypostylid well defined in 12 of 64 specimens with well preserved metalophids, never as prominent as in most specimens of *Perognathoides kleinfelderi.*

Longitudinal ridge originating at centre of metalophid of P_4; ridge appearing only in *Proheteromys floridanus* (Black, 1963a, fig. 4i, k) among other described heteromyids; ridge developed only as short projection from metalophid in ROM 6763 (Figs. 69B, 72C); in ROM 6789 (Fig. 69C) extending to point between protoconid and protostylid; and in holotype (Figs. 69A, 72B), extending anterior to that point, forming anteroconid.

Protolophid with two main cusps, protoconid and protostylid; anteroconid often present; anteroconid at anterior end of longitudinal ridge (Figs. 69A, 72C);

anteroconid not reported in described species of *Perognathus* except modern *P. spinatus peninsulatus*.

In upper premolars, metaloph composed of three main cusps, metacone, hypocone, and hypostyle; hypostyle varying in degree of inflection, more inflected in ROM 6850 (Fig. 69D) than in ROM 6827 (Fig. 69E); protoloph composed of oval protocone, axis directed slightly posterobuccad.

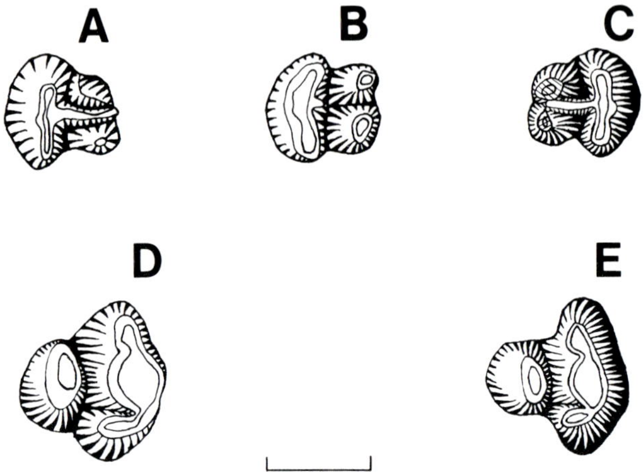

Fig. 69 *Perognathus saskatchewanensis*, premolars in occlusal aspect. Buccal to top of page. Scale 1 mm.
A. LP_4, ROM 6777 (holotype), anterior to right.
B. LP_4, ROM 6763, anterior to right.
C. RP_4, ROM 6789, anterior to left.
D. LP^4, ROM 6850, anterior to left.
E. LP^4, ROM 6827, anterior to left.

MEASUREMENTS

P_4, *AP(61), mean 1.13 ± .01, OR 0.92–1.34, CV 7.1; *BL pro (65), mean 0.87 ± .01, OR 0.76–1.04, CV 8.0; *BL met (62), mean 1.05 ± .01, OR 0.92–1.25, CV 5.7. P^4, *AP(80), mean 1.55 ± .02, OR 1.19–1.95, CV 11.6; **AP(77), mean 1.24 ± .01, OR 1.03–1.45, CV 6.5; *BL pro (82), mean 0.78 ± .01, OR 0.62–0.89, CV 7.7; *BL met (80), mean 1.39 ± .01, OR 1.23–1.61, CV 7.2.

*Measurements taken parallel to occlusal surface.
**Measurements taken perpendicular to posterior wall of metaloph.

DISCUSSION

By definition, the protoloph of P^4 is composed of only one cusp. Although James (1963, p. 107 and fig. 43) showed that a buccal accessory cuspule may be present as a rare variant in *Perognathus*, my separation of upper premolars of *Perognathus saskatchewanensis* from those of *Perognathoides kleinfelderi* has been based on the arbitrary assumption that all specimens lacking any indication of a buccal cuspule could be referred to *Perognathus*. A groove on the posterolingual border of the protoloph was taken to indicate an incipient buccal cuspule. The only exceptions to this criterion were a few very large but worn specimens which were assigned to *Perognathoides*, e.g., ROM 6978 (Fig. 70K).

Perognathus saskatchewanensis is separable from other described species of *Perognathus* on the characteristics of P_4. The population from the Wood Mountain Fauna shows great variability in P_4, but P_4 always has an anteroconid or a central longitudinal ridge, or both.

P. saskatchewanensis is separable from *Perognathoides kleinfelderi* in its smaller average size, less frequent occurrence of a well-defined hypostylid in P_4, occurrence of the anteroconid and longitudinal ridge P_4, and absence of a buccal accessory cuspule in the protoloph of P_4.

Perognathoides kleinfelderi **Storer, 1970a**

Figs. 70; 72D, E

HOLOTYPE. LP_4 (ROM 6927).

REFERRED SPECIMENS. LP_4 (ROM 6907–6926, 6928–6933), RP_4 (ROM 6935–6943, 6945–6960), R. mandibular fragment with RP_4, RM_1 (ROM 6962), LP^4 (ROM 6963–7002, 7063), RP^4 (ROM 7003–7043, 7072).

TENTATIVELY REFERRED. LP_4 (ROM 7309).

TYPE LOCALITY. Kleinfelder Farm Locality, Lsd. 16, Sect. 31, Twp 2, Rge 30, W. 2nd Merid., near Rockglen, Saskatchewan.

DIAGNOSIS. Size comparable to *Perognathoides tertius* and *P. quartus*, larger than *P. cuyamensis*; P_4 almost always showing anteroconid and strong hypostylid; anteroconid strong when present, almost completely separate from protoconid.

DESCRIPTION

Metalophid of P_4 consisting of two major cusps, entoconid and hypoconid, separated by deep anteroposterior valley; low posterior median cuspule on little-worn specimens, shown in ROM 6936 (Fig. 70A); lingual cuspule and hypostylid usually present on metalophid; hypostylid almost always strong, ranging from buccal ridge (Fig. 70B, C), to small but distinct cuspule (Fig. 70D), to large cuspule separated from hypoconid by shallow valley (Fig. 70D), to large cuspule separated from hypoconid by shallow valley (Fig. 70A), and to large cusp, strongly isolated from rest of metalophid (Figs. 70E, 72D), hypostylid observed in 22 of 52 little-worn specimens; in one specimen (Fig. 70F) cuspule anterior to entoconid.

Protolophid of P_4 showing great variability (Fig. 70A–G); anteroconid present in all specimens except ROM 6943 (Fig. 70B); anteroconid usually expressed as ridge connecting anterior ends of protoconid and protostylid, as in ROM 6936, 6945, 6950, and 6927 (Figs. 70A, C, D, E; 72D); in little-worn specimens (e.g., holotype) almost completely separate from protostylid. Rather than spur of protostylid as in *P. cuyamensis* (Wood, 1937, fig. 3); in ROM 6910 (Fig. 70G), anteroconid occupying position farther buccal than usual, bearing small, low, lingual cuspule, more anterior connection of protoconid and protostylid made by anterolingual arm of protoconid.

Posterior connection between protoconid and protostylid usually formed as follows: protoconid, in little-worn specimens, extending posterobuccad toward centre of metalophid, either connecting with metalophid (Figs. 70E, G; 72D), or having posterior end close to metalophid (Fig. 70B); connection of protoconid and metalophid in ROM 6936 (Fig. 70A) effected by short longitudinal ridge running from anterior end of hypoconid to point posterolingual to centre of protoconid. Protostylid joining posterior arm of protoconid in variable manner; in ROM 6945 and 6950 (Fig. 70C, D), simple junction; in ROM 6957 (Fig. 70F) protostylid united with protoconid proper; in ROM 6943 (Fig. 70B) protostylid

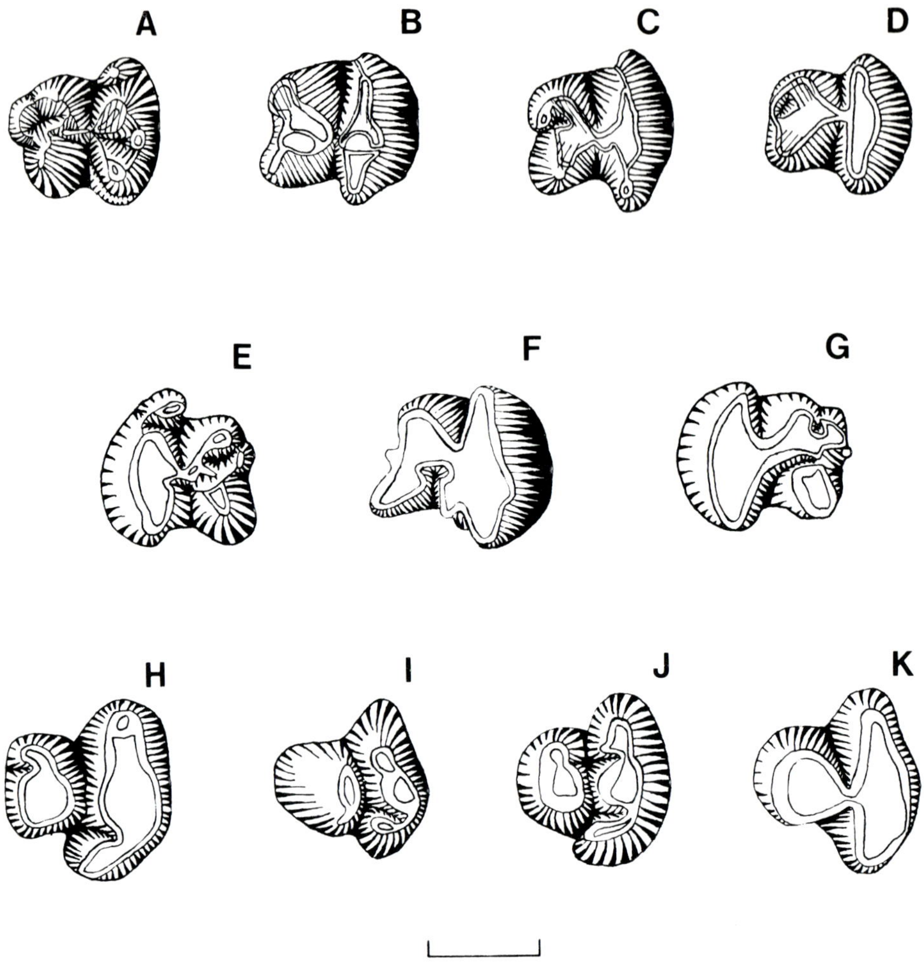

Fig. 70 *Perognathoides kleinfelderi*, premolars in occlusal aspect. Buccal to top of page. Scale 1 mm.
- A. RP_4, ROM 6936, anterior to left.
- B. RP_4, ROM 6943, anterior to left.
- C. RP_4, ROM 6945, anterior to left.
- D. RP_4, ROM 6950, anterior to left.
- E. LP_4, ROM 6927 (holotype), anterior to right.
- F. RP_4, ROM 6957, anterior to left.
- G. LP_4, ROM 6910, anterior to right.
- H. LP^4, ROM 6972, anterior to left.
- I. LP^4, ROM 6983, anterior to left.
- J. LP^4, ROM 6984, anterior to left.
- K. LP^4, ROM 6978, anterior to left.

joined with posterior arm of protoconid at two points, forming small isolated lake; in ROM 6927 (Figs. 70E, 72D), two spurs from posterior arm of protoconid reaching toward protostylid.

In most specimens, as in ROM 6950 (Fig. 70D), small lake formed, bounded by protostylid, anteroconid, and protoconid.

Upper premolar with moderately (Fig. 70H) or greatly (Fig. 70I) inflected hypostyle; by definition, upper premolar of *Perognathoides* with buccal accessory cuspule in protoloph; buccal cuspule of protoloph generally strong (Figs. 70H, J; 72E) but may be weak, disappearing with moderate wear (Fig. 70I, K); connection of the protocone and metaloph always lingual to hypocone.

MEASUREMENTS

P_4, *AP(42), mean 1.42 ± .01, OR 1.17–1.67, CV 8.5; *BL pro (49), mean 1.07 ± .01, OR 0.92–1.23, CV 6.5; *BL met (41), mean 1.30 ± .01, OR 1.12–1.43, CV 6.9. P^4, *AP(70), mean 1.78 ± .03, OR 1.41–2.42, CV 12.4; **AP(68), mean 1.35 ± .02, OR 1.12–1.67, CV 10.4; *BL pro (76), mean 0.89 ± .01, OR 0.75–1.12, CV 10.1; *BL met (75), mean 1.53 ± .01, OR 1.29–1.83, CV 8.5.

*Measurements taken parallel to occlusal surface.
**Measurements taken perpendicular to posterior wall of metaloph.

DISCUSSION

It is clear from the above description that three main factors determine the characteristic forms of P_4 in *P. kleinfelderi*: the shape and degree of isolation of the hypostylid, the size and manner of connection of the anteroconid, and the manner of connection of the protostylid to the posterior arm of the protoconid. These factors combine to produce extreme variability of occlusal pattern (Fig. 70A–G).

Although the sample described here as *P. kleinfelderi* includes specimens of P_4 very similar to Wood's (1937) *P. cuyamensis*, their consistently larger size and more nearly separate anteroconid distinguish them from that species. Likewise, Klingener's specimens (1968, p. 68) of "*Perognathoides* cf. *P. cuyamensis*" from the Norden Bridge local fauna, UMMP 53023, 53026, and 53039, show relatively well-separated anteroconids and greater size than the type material of *P. cuyamensis* and are probably referable to *P. kleinfelderi*.

Peridiomys borealis Storer, 1970a

Figs. 71; 72F–H

HOLOTYPE. LP_4 (ROM 7048)

REFERRED SPECIMENS. LP_4 (ROM 7044–7047), RP_4 (ROM 7049–7052), LP^4 (ROM 7054–7056, 7058–7062, 7064, 7066, 7067), RP^4 (ROM 7068–7071, 7073–7076).

TYPE LOCALITY. Kleinfelder Farm Locality, Lsd. 16, Sect. 31, Twp 2, Rge 30, W. 2nd Merid., near Rockglen, Saskatchewan.

DIAGNOSIS. Heteromyid the size of *Peridiomys kelloggi*, with anteroconid fused to protostylid, and hypostylid variably developed.

DESCRIPTION

Teeth of about same size as those of *Peridiomys kelloggi*, somewhat larger than those of *P. oregonensis*; the holotype (Figs. 71A, 72F) with two major cusps in metalophid plus central posterior cuspule, no posterior cuspule visible in ROM 7051 (Fig. 72B), but bulge in metalophid near junction of entoconid and hypoconid may indicate cuspule.

Anteroconid present in P_4, usually dominated by protostylid (Figs. 71A, B; 72F); never completely connected to protoconid in little-worn teeth, unlike

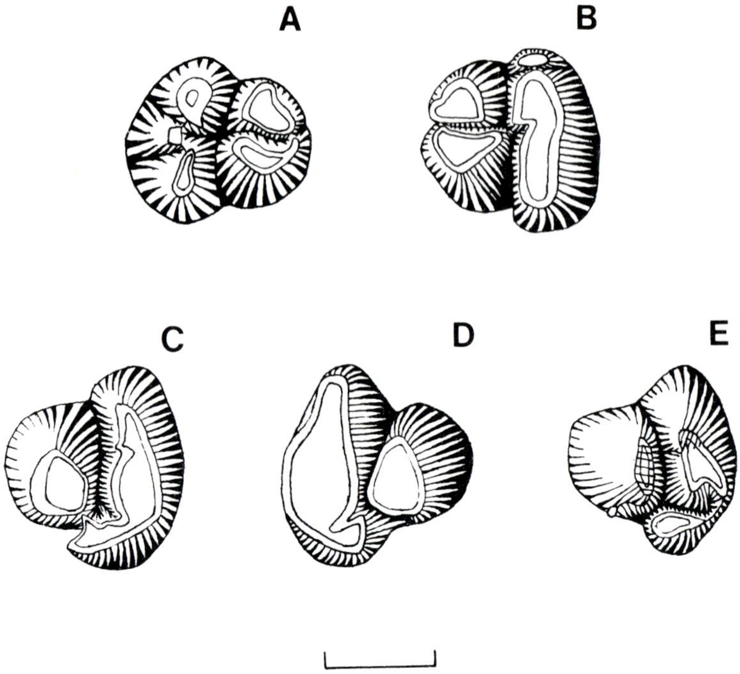

Fig. 71 *Peridiomys borealis*, premolars in occlusal aspect. Buccal to top of page. Scale 1 mm.
A. LP_4, ROM 7048 (holotype), anterior to right.
B. RP_4, ROM 7051, anterior to left.
C. LP^4, ROM 7067, anterior to left.
D. RP^4, ROM 7069, anterior to right.
E. LP^4, ROM 7058, anterior to left.

situation in *P. kelloggi* (Wood, 1936a, fig. 6); protoconid never significantly larger than protostylid as in holotype of *P. rusticus* (Wood, 1935, fig. 97).

Large hypostylid in some specimens (Fig. 71B) as in holotype of *P. rusticus*, not appearing in holotype (Figs. 71A, 72F); junction of lophs in typical heteromyine manner, at lingual and buccal sides of crown.

Upper premolars often characterized by short spur on lingual side of hypostyle; spur may be prominent (Figs. 71C, 72G), weak (Figs. 71D, 72H), or absent (Fig. 71E); low lingual cuspule may be present on protoloph (Fig. 71E); shape of protoloph on many specimens (Figs. 71C, D; 72G, H) suggests presence of more than one cusp; protoloph and metaloph usually uniting at tip of hypostyle (Figs. 71C, E; 72G), sometimes joining closer to centre of metaloph (Figs. 71D, 72H).

MEASUREMENTS

P_4, *AP(9), mean 1.45 ± .05, OR 1.25–1.67 , CV 9.7; *BL pro (9), mean 1.14 ± .04, OR 0.99–1.32, CV 9.7; *BL met (9), mean 1.32 ± .04, OR 1.19–1.58, CV 9.8. P^4, *AP(18), mean 1.97 ± .04, OR 1.67–2.28, CV 8.1; ** AP(18), mean 1.55 ± .03, OR 1.32–1.72, CV 7.1; *BL pro (19), mean 1.02 ± .03, OR 0.79–1.16, CV 10.8; *BL met (18), mean 1.71 ± .02, OR 1.56–1.85, CV 5.8.

*Measurements taken parallel to occlusal surface.
**Measurements taken perpendicular to posterior wall of metaloph.

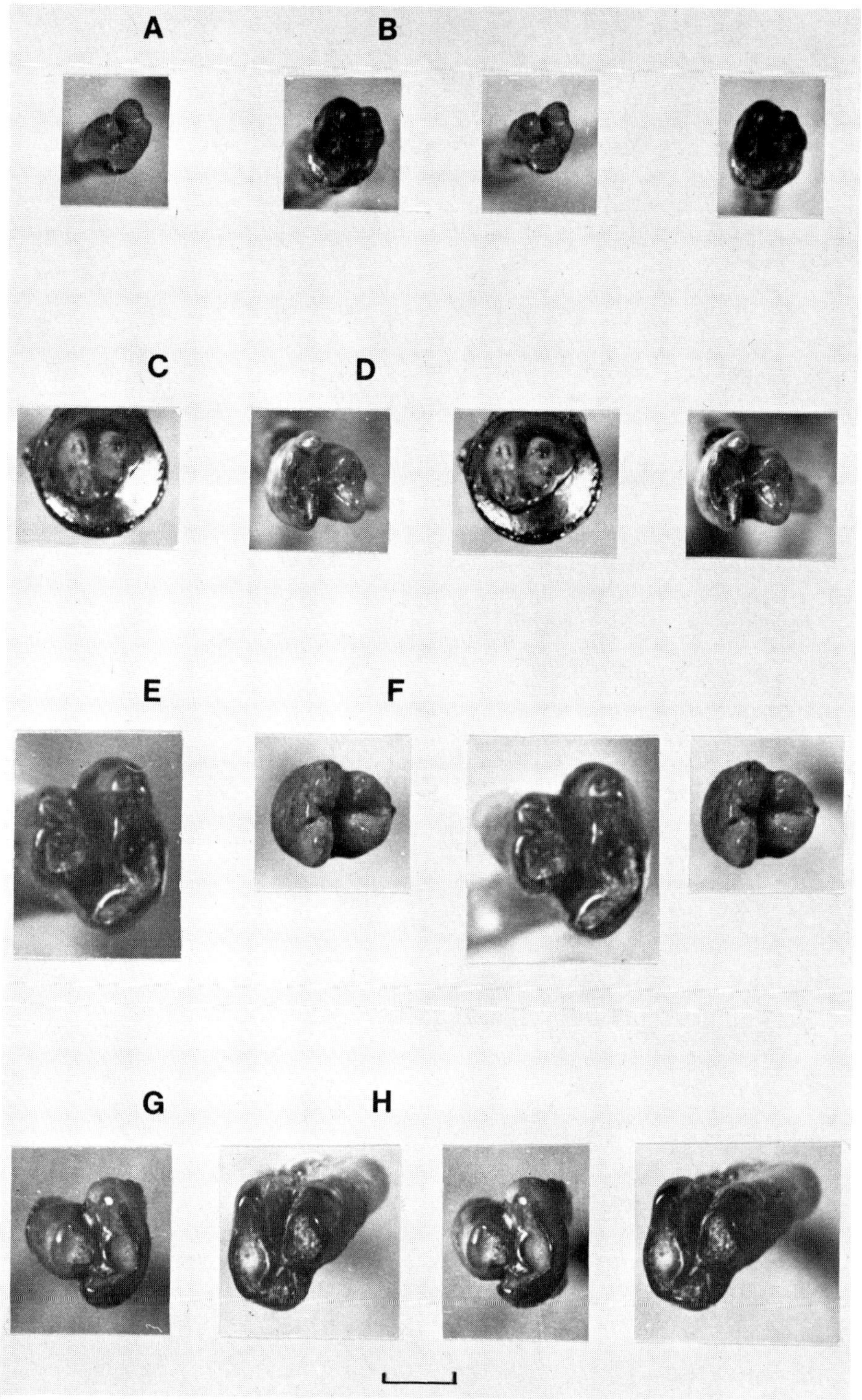

Fig. 72 *Mookomys*, *Perognathus*, *Perognathoides*, and *Peridiomys*, premolars in occlusal aspect. Buccal to top of page. Scale 1 mm.

A. *Mookomys* sp., LP^4, ROM 7086, anterior to left.
B,C. *Perognathus saskatchewanensis,* LP_4, anterior to right.
B. ROM 6777 (holotype).
C. ROM 6763.
D,E. *Perognathoides kleinfelderi.*
D. LP_4, ROM 6927 (holotype), anterior to right.
E. LP^4, ROM 6984, anterior to left.
F-H. *Peridiomys borealis.*
F. LP_4, ROM 7048 (holotype), anterior to right.
G. LP^4, ROM 7067, anterior to left.
H. RP^4, ROM 7069, anterior to right.

DISCUSSION

P. borealis is distinguished from other *Peridiomys* by the weak anteroconid on P_4 and by the spur of the hypostyle on P^4. Since wear may obliterate the features of the anteroconid on P_4, and the spur of the hypostyle of P^4 is a variable feature, individual specimens of this species may not always be recognizable.

Diprionomys cf. *D. agrarius* Wood, 1935

Fig. 73

REFERRED SPECIMENS. LP_4 (ROM 7077, 7078), RP_4 (ROM 6805, 6815, 7079), LP^4 (ROM 7057, 7065, 7080, 7081), RP^4 (ROM 7082–7084).

DESCRIPTION

Lower premolars with three cusps in protolophid, arranged in semicircle; deep anteroposterior valley in protolophid; three cusps in metalophid; hypostylid prominent ridge in ROM 6805 (Fig. 73C); connection of lophs between protostylid and hypoconid; protoconid and entoconid not joining.

Connection of lophs in upper premolar (not figured) between protocone and hypostyle.

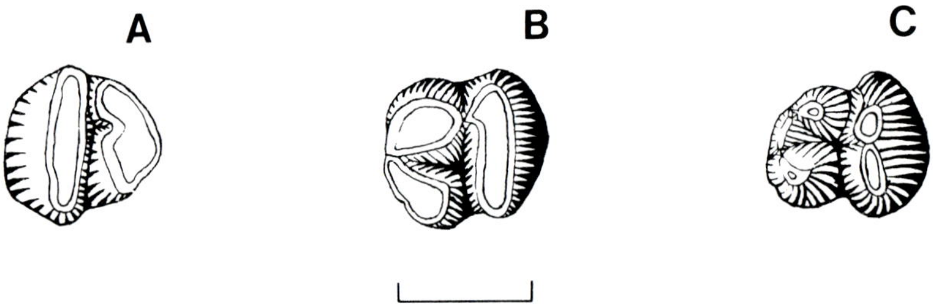

Fig. 73 *Diprionomys* cf. *D. agrarius*, lower premolars in occlusal aspect. Buccal to top of page. Scale 1 mm.
A. LP_4, ROM 7078, anterior to right.
B. RP_4, ROM 7079, anterior to left.
C. RP_4, ROM 6805, anterior to left.

MEASUREMENTS

P_4, AP(5), mean 1.14 ± .01, OR 1.10–1.19, CV 2.9; BL pro (5), mean 0.89 ± .02, OR 0.86–0.96, CV 4.5; BL met (5), mean 1.03 ± .02, OR 0.97–1 08, CV 3.9. P^4, AP(7), mean 1.60 ± .04, OR 1.41–1.76, CV 6.9; BL pro (7), mean 0.79 ± .01, OR 0.75–0.84, CV 3.8; BL met (7), mean 1.28 ± .03, OR 1.14–1.41, CV 7.0.

DISCUSSION

Connection of P_4 lophs between the protostylid and hypoconid, with the protoconid and entoconid not joining, are diagnostic of *Diprionomys* (Wood, 1935). Lower premolars resemble the holotype of *D. agrarius* (Wood, 1935, fig. 103) in pattern, but are slightly smaller. Similarities are also shown to *D* cf. *D. agrarius* from the Norden Bridge local fauna (Klingener, 1968, fig. 20).

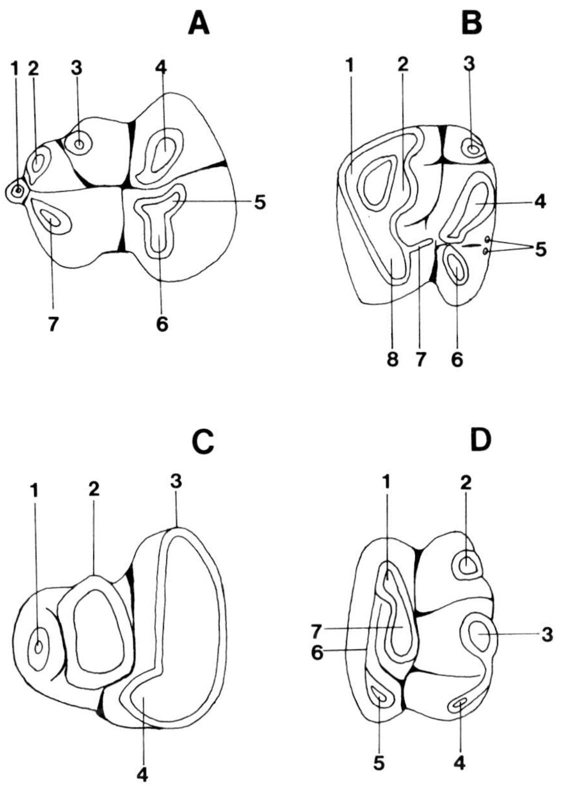

Fig. 74 *Lignimus montis*. Terminology used in description of cheek teeth. Anterior to left, buccal to top of page. Not to scale.

A. RP$_4$: 1,2, cusps of anterior cingulum; 3, protostylid; 4, hypostylid; 5, hypoconid; 6, entoconid; 7, protoconid.

B. RM$_1$ or M$_2$: 1, protostylid; 2, protoconid; 3, hypostylid; 4, hypoconid; 5, pair of cuspules at posterior end of valley between entoconid and hypoconid; 6, entoconid; 7, spur of metaconid; 8, metaconid.

C. LP^4: 1, ?anterocone; 2, protoloph; 3, metaloph; 4, hypostyle.

D. LM^1 or M^2: 1, paracone; 2, metacone; 3, hypocone; 4, hypostyle; 5, protostyle; 6, anterior cingulum; 7, protocone.

Family Geomyidae
Subfamily Entoptychinae

Lignimus Storer, 1970a

TYPE SPECIES. *Lignimus montis* Storer, 1970a.

REFERRED. *Lignimus hibbardi* Storer, 1973.

HORIZONS. Barstovian of Saskatchewan; Clarendonian of Kansas, Nebraska.

DIAGNOSIS OF GENUS. (Storer, 1973). Teeth high-crowned but rooted. Highly worn teeth develop interruptions in enamel in some species. Lower molars with strong X-pattern, protostylid forming lake in metalophid. Upper premolars with anterocones variably present. Central enamel lakes formed in worn upper molars, anterior cingula prominent in upper molars.

Lignimus montis Storer, 1970a
Figs. 74–76

HOLOTYPE. LP_4 (ROM 7108).

PARATYPES. LP_4 (ROM 7109), RM_1 or M_2 (ROM 7140, 7145), RP^4 (ROM 7158, 7159), LM^1 or M^2 (ROM 7167).

REFERRED SPECIMENS. RP_4 (ROM 6944, 7053, 7110), LM_1 or M_2 (ROM 7111–7131, 7150, 7151), RM_1 or M_2 (ROM 7132–7139, 7141–7144, 7146, 7147, 7153, 7155), LM_3 (ROM 7148, 7149, 7152), RM_3 (ROM 7154, 7156), LP^4 (ROM 7157), LM^1 or M^2 (ROM 7160–7166, 7168, 7182), RM^1 or M^2 (ROM 7169–7175, 7177–7180, 7185, 7188), LM^3 (ROM 7181, 7183), RM^3 (ROM 7176, 7184, 7186, 7187, 7189).

TYPE LOCALITY. Kleinfelder Farm Locality, Lsd. 16, Sect. 31, Twp 2, Rge 30, W. 2nd Merid., near Rockglen, Saskatchewan.

DIAGNOSIS OF SPECIES. Small entoptychine; highly-worn teeth never developing interruptions in enamel as in *Entoptychus*. P_4 with prominent protoconid; cusps of anterior cingulum large, separated from each other and from protoconid by deep valleys; pit formed by junction of lophs in centre of well-worn P_4. Lower molars showing little tendency to form central lakes; lake formed labial to centre of crown in well-worn M_{1-2}; lower molars with strong X-pattern, emphasized by anteriorly directed V-shaped ridge connecting entoconid and hypoconid. Protoloph of P^4 proportionally narrower than other entoptychines, rounded in well-worn P^4, uniting first with hypostyle of metaloph without forming central lake. Protocone of upper molars large, high, somewhat isolated.

DESCRIPTION

In holotype P_4, ROM 7108 (Figs. 75A, 76A), hypoconid and entoconid fused, buccal cuspule present on metalophid; large protoconid in protolophid; protostylid as high, almost as wide as protoconid, not as long anteroposteriorly; two additional cuspules present, both prominent, anterior one smaller and lower than posterior; protostylid and cingular cuspules separated from each other and from protoconid by deep valleys; in paratype ROM 7109 (Figs. 75B, 76B), union of protolophid and metalophid forming central lake, protostylid and anterior cingulum joined to protoconid.

Metalophid of lower molars (Figs. 75C, D; 76C, D) made up of two large cusps, metaconid and protoconid, plus cingular protostylid surrounding anterobuccal side of protoconid, projecting posterad toward hypostylid; lake formed between protostylid and protoconid (Fig. 75E).

Lower molars characterized by strong X-pattern of cusps and lophs (Figs. 75C–G; 76C, D); in M_1 and M_2, anteriorly directed V-shaped ridge connecting hypoconid and entoconid (Figs. 75C–E; 76C, D); V-shaped ridge joining either with metaconid (Figs. 75C, 76C) or with spur projecting from metaconid (Fig. 75E); posteriorly directed valley formed behind V-shaped ridge, sometimes blocked at posterior end by two small cuspules (Figs. 75C, 76C).

Paratypic P^4s (Figs. 75H, I; 76E, F) too worn to show composition of individual lophs; metaloph presumably with at least three cusps; number of cusps in nearly round protoloph cannot be determined; strong anterocone present in ROM 7159, but not in ROM 7158; both specimens showing protoloph in process of uniting with entostyle, forming no central lake in either specimen; P^4 proportionally narrow.

Protoloph of upper molar (Fig. 75J–L) composed of paracone, protocone,

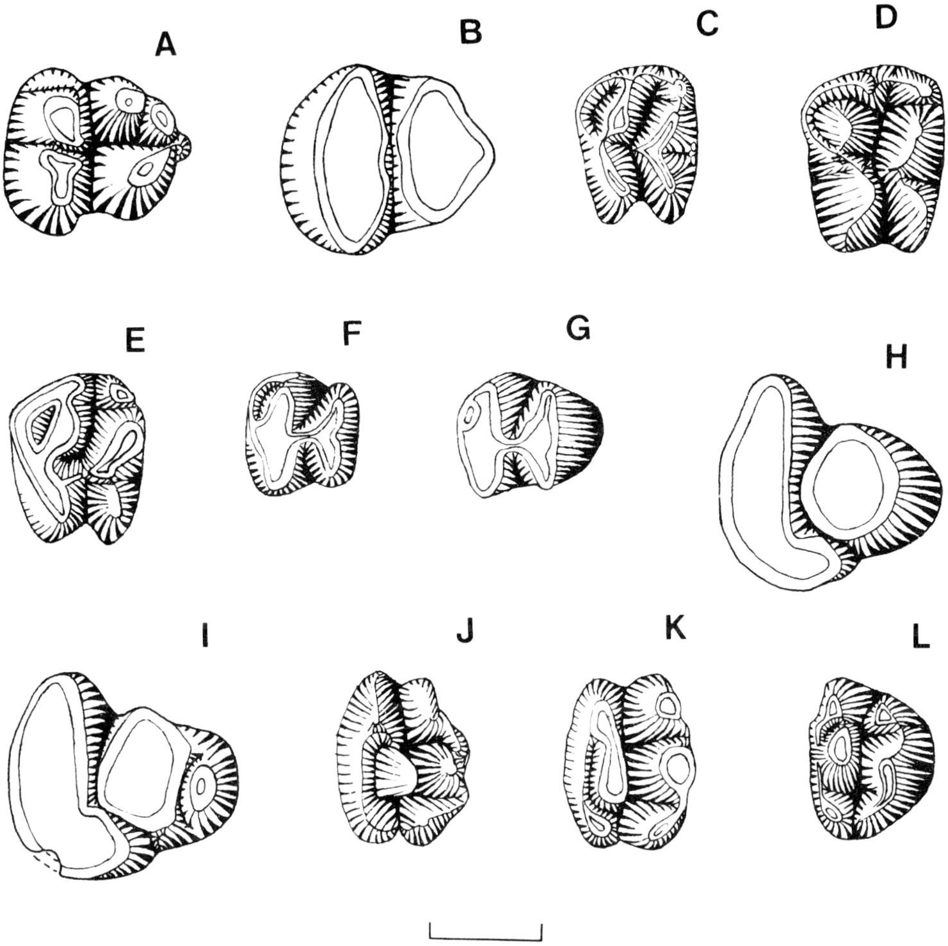

Fig. 75 *Lignimus montis*, cheek teeth in occlusal aspect. Buccal to top of page. Scale 1 mm.

A. LP_4, ROM 7108 (holotype), anterior to right.
B. LP_4, ROM 7109 (paratype), anterior to right.
C. RM_1 or M_2, ROM 7140 (paratype), anterior to left.
D. RM_1 or M_2, ROM 7145 (paratype), anterior to left.
E. RM_1 or M_2, ROM 7133, anterior to left.
F. RM_3, ROM 7154, anterior to left.
G. RM_3, ROM 7156, anterior to left.
H. RP^4, ROM 7158 (paratype), anterior to right.
I. RP^4, ROM 7159 (paratype), anterior to right.
J. LM^1 or M^2, ROM 7167 (paratype), anterior to left.
K. LM^1 or M^2, ROM 7163, anterior to left.
L. LM^3, ROM 7181, anterior to left.

and protostyle; strong anterior cingulum between paracone and protostyle; large, prominent protocone generally separated from anterior cingulum by deep valley (Fig. 75J, K); in ROM 7181 (Fig. 75L), however, anterior face of protocone projecting anterad to break anterior cingulum; first union of protocone with paracone (Fig. 75K); strong, high loph connecting hypocone and hypostyle, weak, lower loph connecting hypocone and metacone; first confluence of lophs between protostyle and hypostyle; in well-worn specimens, series of central lakes formed by progressive junctions of lophs from lingual side to buccal.

Enamel borders of crowns never showing interruptions in enamel; many specimens with uneven borders, slight rise in enamel sometimes occurring at lingual or buccal side; crowns varying greatly in dorsoventral height.

MEASUREMENTS

P_4, AP(3), mean 1.78, OR 1.63–1.98; BL pro (5), mean 1.32 ± .02, OR 1.23–1.59, CV 3.8; BL met (4), mean 1.65, OR 1.57–1.74. M_1 or M_2, AP(37), mean 1.53 ± .03, OR 1.20–1.98, CV 11.8; BL met (36), mean 1.56 ± .02, OR 1.41–1.85, CV 7.1; BL hyp (36), mean 1.53 ± .03, OR 1.17–1.85, CV 9.8. M_3,

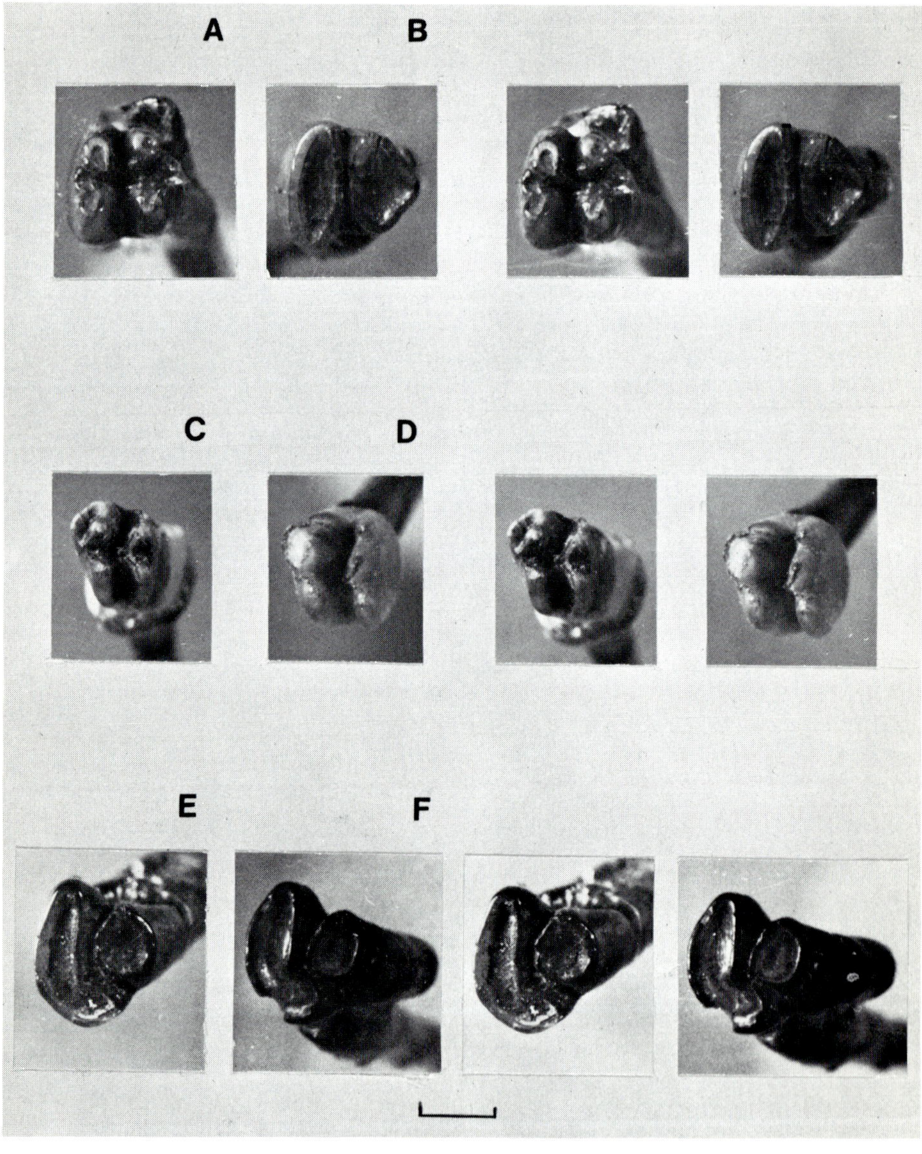

AP(5), mean 1.32 ± .07, OR 1.14–1.54, CV 12.1; BL met (5), mean 1.15 ± .06, OR 1.01–1.41, CV 12.1; BL hyp (5), mean 1.05 ± .03, OR 0.97–1.12, CV 5.7. P^4, AP(2), OR 2.20–2.53; BL pro (2), OR 1.14–1.26; BL met (2), OR 1.87–1.92. M^1 or M^2, AP(24), mean 1.10 ± .02, OR 0.97–1.44, CV 10.9; BL pro (24), mean 1.43 ± .04, OR 1.19–1.89, CV 11.9; BL met (24), mean 1.36 ± .03 , OR 1.16–1.80, CV 12.5. M^3, AP(6), mean 0.97 ± .02, OR 0.88–1.03, CV 9.2; BL pro (6), mean 1.25 ± .04, OR 1.09–1.39, CV 7.2; BL met (6), mean 1.10 ± .04, OR 0.97–1.28, CV 9.1.

DISCUSSION

Lophs of lower premolars of *Lignimus* are united laterally, as in entoptychines, and a central basin is formed. In geomyines *Dikkomys* (Wood, 1936b, fig. 32A; Black, 1961b, fig. 5B) and *Mojavemys* (Lindsay, 1972, fig. 34A), lophs are united centrally, and no basin is formed. Because of these gross differences in structure from geomyine premolars, *Lignimus* cannot be assigned to the Geomyinae.

Lignimus, like the geomyines, shows an X-pattern in lower molars. Buccal connections of the lophs are strong, and strengthening of the central connection may have been a convergent trend. Strength of the central connection varies in *Lignimus montis* (Fig. 76C, D).

Lignimus montis is distinct in many features from previously described entoptychines. Still, *L. montis* is clearly related to Lower Miocene forms, especially *Gregorymys*. *G. curtus* (Wood, 1936b, figs. 12, 13), for example, shows incipient development of the cingular protostylid on the lower molars, and *G. formosus*, *G. curtus*, and *G. riggsi* (Wood, 1936b, figs. 10, 12, 15) have strong anterior cingula on P_4. *Lignimus* is most likely to have been derived from an undescribed species of *Gregorymys*.

A report on *Lignimus* from the Clarendonian of Kansas and Nebraska has been published (Storer, 1973), and the revision of the generic diagnosis is based on the new material. *Lignimus* is likely to be found in more deposits, as screening for microvertebrates is widely applied.

The latest record of an entoptychine is from the Moonstone Formation, Hemphillian, of Wyoming (Love, 1961, p. I-35). But the sample of *L. montis* is the youngest described population, and gives the first indication of trends in the dental evolution of post-Arikareean entoptychines.

Fig. 76 *Lignimus montis*, cheek teeth in occlusal aspect. Buccal to top of page. Scale 1 mm.

A. LP_4, ROM 7108 (holotype), anterior to right.
B. LP_4, ROM 7109 (paratype), anterior to right.
C. RM_1 or M_2, ROM 7140 (paratype), anterior to left.
D. RM_1 or M_2, ROM 7145 (paratype), anterior to left.
E. RP^4, ROM 7158 (paratype), anterior to right.
F. RP^4, ROM 7159 (paratype), anterior to right.

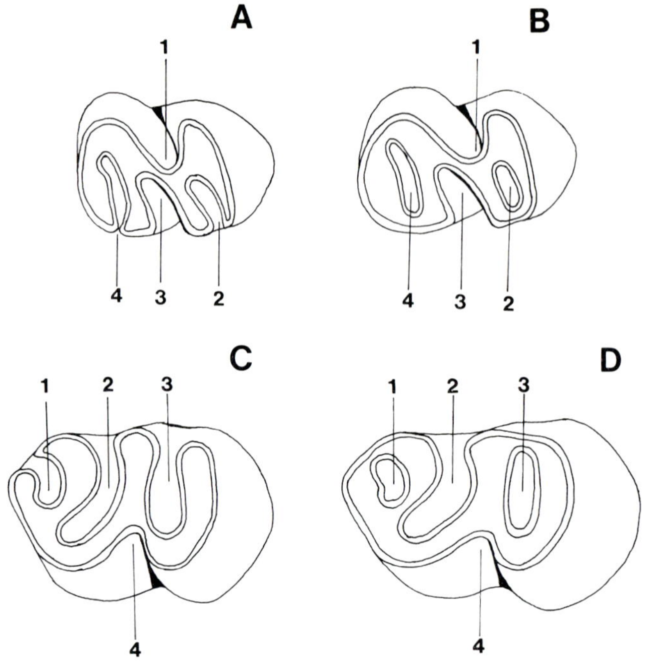

Fig. 77 Castoridae. Terminology used in description of cheek teeth. Anterior to left, buccal to top of page. Not to scale.

A. RP_4 lightly worn with flexids open: 1, hypoflexid; 2, metaflexid; 3, mesoflexid; 4, anteroflexid.

B. RP_4 more heavily worn with anteroflexid and metaflexid closed; 1, hypoflexid; 2, metafossettid; 3, mesoflexid; 4, anterofossettid.

C. LP^4 lightly worn with flexi open: 1, paraflexus; 2, mesoflexus; 3, metaflexus; 4, hypoflexus.

D. LP^4 more heavily worn with paraflexus and metaflexus closed: 1, parafossette; 2, mesoflexus; 3, metafossette; 4, hypoflexus.

Family Castoridae

Monosaulax cf. *M. pansus* (Cope), 1874

Fig. 78

REFERRED SPECIMENS. R. mandibular fragment with RI_1, RP_4, RM_{1-2} (L. Yost specimen; NMC 9393), LP_4 (SMNH P123, P552; ROM 7613), L. mandibular fragment with LP_4, LM_{1-2} (SMNH 7988), RP_4 (SMNH P122—2 specimens, P552—2 specimens, P566; ROM 7614), L. lower molar (SMNH P557; ROM 7253, 7254, 7615), R. lower molar (ROM 7255–7259), RdP^4 (ROM 7260), LP^4 (SMNH P123, P552; ROM 7261, 7616), RP^4 (SMNH P385; ROM 7277), L. maxillary fragment with LP^4, LM^1 (NMC 9012), R. maxillary fragment with RP^4, RM^1 (SMNH P76), LM^1 or LM^2 (ROM 7269–7275, 7617–7619), RM^1 or M^2 (ROM 7262–7268), RM^3 (SMNH P122).

DESCRIPTION

Lower cheek teeth (Fig. 78A, B) with typical *Monosaulax* (*sensu* Stirton) pattern; small round cement lake just anterior to buccal end of anterofossettid

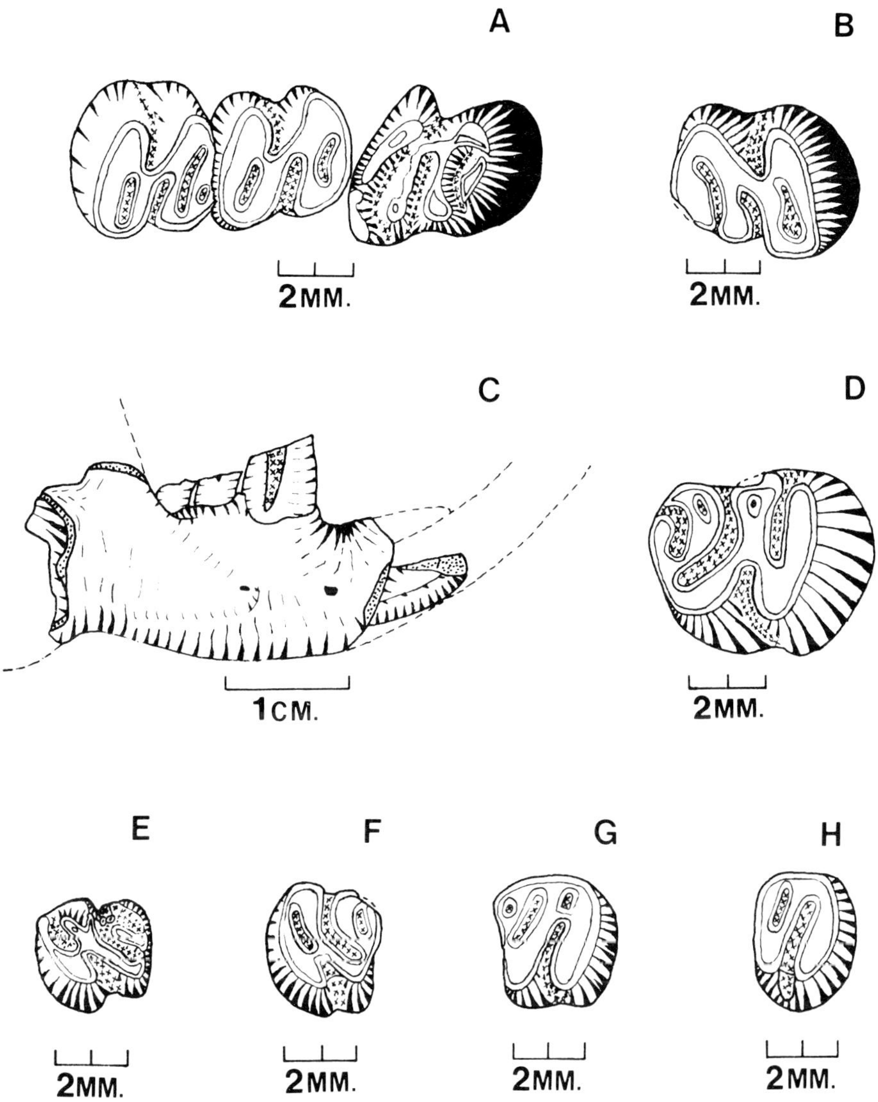

Fig. 78 *Monosaulax* cf. *M. pansus*, cheek teeth and mandible. Scales as noted.

A. LP_4, LM_{1-2}, SMNH 7988, occlusal aspect, anterior to right, buccal to top of page.

B. RP_4, SMNH P566, occlusal aspect, anterior to left, buccal to top of page.

C. R. mandible with RI_1, RP_4, RM_{1-2}, L. Yost specimen, lateral aspect, anterior to right, occlusal to top of page.

D. LP^4, ROM 7261, occlusal aspect, anterior to left, buccal to top of page.

E. RM^1 or M^2, ROM 7265, occlusal aspect, anterior to right, buccal to top of page.

F. RM^1 or M^2, ROM 7266, occlusal aspect, anterior to right, buccal to top of page.

G. LM^1 or M^2, ROM 7273, occlusal aspect, anterior to left, buccal to top of page.

H. LM^1 or M^2, ROM 7270, occlusal aspect, anterior to left, buccal to top of page.

of M_2 (Fig. 78A). Mental foramen about midway between dorsal and ventral surfaces of ramus, in diastematic region just anterior to P_4 (Fig. 78C); ascending ramus arising at posterior end of M_1; coronid process low; ventral border of horizontal ramus converging with line of tooth row anteriorly, curving downward at anterior end of angle; small accessory foramen in some specimens at anterior end of masseteric ridge, beneath posterior end of P_4.

Typical *Monosaulax* patterns seen in P^4 (Fig. 78D) and upper molars (Fig. 78E–H). Specimens in progressively more worn condition including ROM 7265 (Fig. 78E), in which all flexi open; ROM 7266 (Fig. 78F), in which occlusal surface proportionally broader and pattern reduced to short hypoflexus, elongate metafossette with buccal end close to lingual end of hypoflexus, long, curved mesoflexus, and small, oval parafossette; ROM 7273 (Fig. 78G), in which parafossette small and circular, mesoflexus closed and oval, hypoflexus lengthened, and metaflexus shortened; and ROM 7270 (Fig. 78H), worn, in which hypoflexus still long and open, but parafossette and metafossette gone. In ROM 7273, occlusal surface concave anteriorly, wear bringing occlusal surface to part of column with different cross-sectional shape from that seen in little-worn tooth; occlusal surface of ROM 7270 proportionally wide and short, anterior face flattened; in all upper molars, entire occlusal surface surrounded by enamel, change in shape of occlusal surface occurring because of different cross sections at different heights of column.

MEASUREMENTS

P_4, *AP(8), mean 5.1 ± .1, OR 4.7–5.5, CV 5.9; *BL met (11), mean 4.2 ± .1, OR 3.4–5.1, CV 9.5. Lower molars, *AP(14), mean 2.9 ± .1, OR 2.7–3.5, CV 10.3; **AP(14), mean 3.3 ± .1, OR 2.8–4.2, CV 12.1; *BL(15), mean 3.7 ± .1, OR 3.2–4.6, CV 10.8; **BL(14), mean 3.7 ± .1, OR 3.1–4.5, CV 10.8. dP^4, *AP(1), 3.2; *BL(1), 3.4. P^4, *AP(7), mean 4.4 ± .0, OR 4.2–4.6, CV 2.3; *BL(7), mean 4.5 ±.1, OR 3.9–5.0, CV 8.9. M^1 or M^2, *AP(16), mean 2.8 ± .1, OR 2.4–3.2, CV 7.1; **AP(18), mean 3.0 ± .0, OR 2.6–3.3, CV 6.7; *BL(17), mean 3.2 ± .1, OR 2.7–4.1, CV 12.5; **BL(18), mean 3.1 ± .1, OR 2.6–3.9, CV 12.9. M^3, **AP(1), 2.7; **BL(1), 3.5.

*Measurements taken on column of tooth, not necessarily parallel to occlusal surface.

**Measurements of occlusal surface only.

DISCUSSION

The Wood Mountain specimens described above show a typical *Monosaulax* pattern (*sensu* Stirton, 1935), as the anterostriids (grooves on the buccal aspects beneath the anteroflexids) of lower cheek teeth are extremely shallow (Fig. 78A) and are seen only in unworn specimens.

A small lake anterior to the buccal end of the anterofossettid is also seen in *Monosaulax* sp. from the Flint Creek Fauna (Black, 1961a, fig. 2).

Teeth and ramal fragments from the Wood Mountain Fauna resemble *Monosaulax pansus* in visible morphological characters (Stirton, 1935, figs. 66–73). Teeth from Wood Mountain may be slightly smaller than those in Stirton's 1935 sample. Stirton, however, gave no measurements of teeth, and I have not seen his specimens. As Stirton (1935, p. 420) pointed out, species of *Monosaulax* "are not very clearly distinguishable".

As can be seen in the measurements, differences in cross-section of the crown at different heights result in high coefficients of variation, higher in the upper and lower molars than in the premolars.

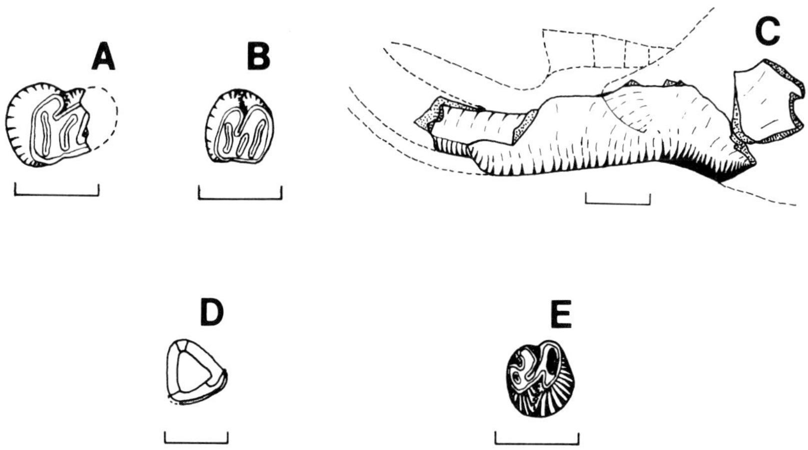

Fig. 79 *Amblycastor* sp., cheek teeth, mandibular fragment, and incisor. Scales 1 cm.

A. LP_4, NMC 9387, occlusal aspect, anterior to right, buccal to top of page.

B. LM_1 or M_2, NMC 9387, occlusal aspect, anterior to left, buccal to top of page.

C. L. mandibular fragment with LI_1, NMC 9379, lateral aspect, anterior to left, occlusal to top of page.

D. LI_1, same specimen as in C, anterior aspect of cross section, dorsal to top of page.

E. LP^4, SMNH P256, occlusal aspect, anterior to left, buccal to top of page.

Amblycastor sp.
Fig. 79

REFERRED SPECIMENS. L. mandibular fragment with LI_1 (NMC 9379), LI_1 (NMC 9387), LP_4 (NMC 9387), LM_1 or M_2 (NMC 9387; ROM 7276), LP^4 (SMNH P236).

DESCRIPTION

Specimens representing *Amblycastor*-like rodent; resembling, but much smaller than, corresponding elements of *A. fluminis* (Stirton, 1935, figs. 47–52); lower premolar (Fig. 79A) not resembling lower premolar of *Hystricops* (Stirton, 1935, fig. 60) in shape or pattern; showing robust appearance of *Amblycastor* (Stirton, 1935, fig. 53). Lower molars (Fig. 79B) resembling both *Amblycastor* and *Hystricops* in simple pattern.

Fragment of mandibular ramus (Fig. 79C) broken so that mental foramen and alveoli missing, angle poorly represented; shape of ventral surface of jaw similar to that of *Amblycastor fluminis* (Stirton, 1935, fig. 50), but specimen smaller; broken lower incisor figured in cross section (Fig. 79D).

Upper premolar (Fig. 79E) less worn than specimen figured by Stirton (1935, fig. 51), similar to it in shape; complex pattern of lakes anterior to mesofossette not present.

MEASUREMENTS

I_1, AP(1), 9.2; transverse (1), 8.2. P_4, BL met (1), 9 1. M_1 or M_2, AP(2), OR 6.5–7.3; BL met (2), OR 8.2–8.7; BL hyp (2), OR 8.0–8.5. P^4, AP(1), 7.0; BL pro (1), 7.0; BL met (1), 8.3.

DISCUSSION

This rare form from the Wood Mountain Fauna is not sufficiently represented for specific description. The presence of the genus increases similarity to the Lower Snake Creek Fauna.

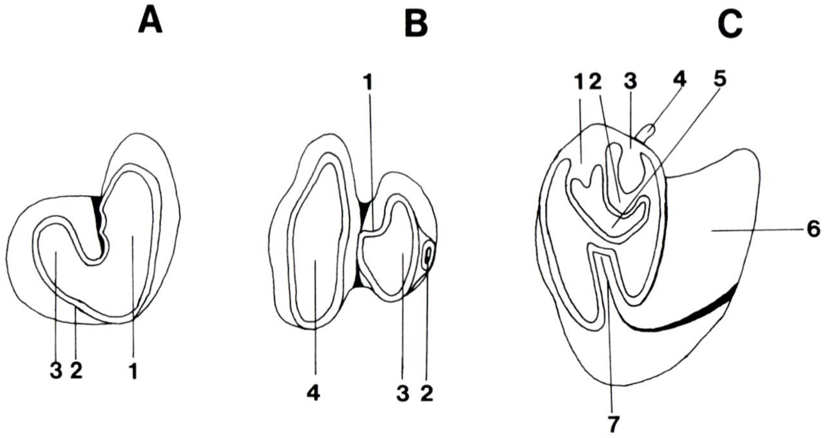

Fig. 80 Ochotonidae. Terminology used in description of cheek teeth. Anterior to left, buccal to top of page. Not to scale.

A. RP_3: 1, talonid; 2, anterolingual fold; 3, trigonid.

B. RP_4 or M_1 or M_2: 1, anterobuccal wall of talonid; 2, posterolophid; 3, talonid; 4, trigonid.

C. LP^4: 1, anterobuccal cuspule; 2, amphicone; 3, posterobuccal cuspule; 4, posterobuccal root; 5, cement-filled crescentic valley; 6, lingual root; 7, lingual hypostria.

Order Lagomorpha
Family Ochotonidae

Hesperolagomys fluviatilis Storer, 1970a
Figs. 81, 82

HOLOTYPE. L. lower molariform tooth (ROM 7320).

PARATYPES. RP_3 (ROM 7306), ?LP^4 (ROM 7345), ?RP^4 (ROM 7352), ?LM^1 or M^2 (ROM 7343, 7348).

REFERRED SPECIMENS. LP_3 (ROM 7302–7304), RP_3 (ROM 7305, 7307), LdP_4 (ROM 7308), L. lower molariform tooth (ROM 7310–7319, 7321–7324), R. lower molariform tooth (ROM 7325–7338, 7416), RP^3 (ROM 7339), RdP^4 (ROM 7340), ?LP^4 (ROM 7351), ?RP^4 (ROM 7354), ?LM^1 or M^2 (ROM 7344, 7346, 7347, 7349, 7350), ?RM^1 or M^2 (ROM 7353).

TYPE LOCALITY. Kleinfelder Farm Locality, Lsd. 16, Sect. 31, Twp 2, Rge 30, W. 2nd Merid., near Rockglen, Saskatchewan.

DIAGNOSIS. Size of teeth similar to those of *Hesperolagomys galbreathi*. Anterolingual fold of P_3 much weaker than in *H. galbreathi*; anterior extension of talonid in lower molariform teeth more prominent than in *H. galbreathi*; posterobuccal angle of upper molariform teeth more reduced than in *H. galbreathi*.

DESCRIPTION

Lower cheek teeth high-crowned but rooted; figured paratype P_3 (Figs. 81A, 82A) not showing strong anterolingual fold of *Hesperolagomys galbreathi* (Clark, Dawson, and Wood, 1964, fig. 4A), some other specimens with very weakly developed folds; little-worn specimen, ROM 7306, with series of plications on anterior face of talonid; no cement on lingual surface of crown of P_3.

Deciduous fourth lower premolar (Fig. 81B) showing greater complexity than in dP_4 of *Oreolagus nebrascensis* figured by Dawson (1965, fig. 29); in trigonid, anterolingual corner strongly developed, with lingually oriented anterolingual valley; anterior extension of talonid far stronger than in *O. nebrascensis*; narrow posterolophid posterior to talonid.

Lower molariform teeth with talonids longer than (Figs. 81C–E; 82B) or as long as (Fig. 81F, G) trigonids; shape and strength of anterior projection of talonid varying, but always stronger than in *H. galbreathi*; in almost all specimens, anterobuccal wall of talonid describing smooth curve (Figs. 81D–F; 82B) or obtuse angle (Fig. 81C), but one specimen (Fig. 81G) showing plications resembling those seen in *Russellagus vonhofi* (Figs. 83D, E; 85B); little worn specimens (Fig. 81E, G) showing narrow posterolophids as in most leporids, *Russellagus*, and perhaps *Oreolagus wallacei* (Dawson, 1965, fig. 31).

Deciduous upper premolar, probably RdP^4 (Fig. 81H, I), resembling permanent upper molariform teeth in pattern, but showing almost unreduced posterobuccal border; crown short dorsoventrally; ventral border almost parallel with surface of crown; three subequal roots, one lingual and two buccal; much smaller than permanent molariform teeth, as in *Hesperolagomys galbreathi* (Clark, Dawson, and Wood, 1964, pp. 35–36).

Upper third premolar, ROM 7339 (Fig 81J), very similar to corresponding teeth of ?*Desmatolagus schizopetrus* (Dawson, 1965, fig. 15) and *Russellagus vonhofi* (Figs. 83H, I; 84D, E).

Permanent upper molariform teeth high-crowned but rooted (Figs. 81K, 82C); two short buccal roots and one large lingual root; upper molariform teeth separable into distinct groups, P^4 and M^1 or M^2.

Probable P^4 (Figs. 81L, M; 82D, E) showing narrow lingual hypostriae with nearly parallel borders; posterobuccal corners far more reduced than those in either *H. galbreathi* (Clark et al., 1964, fig. 3D) or *Russellagus vonhofi* (Figs 83M, O; 84G), reduced to lesser degree than those of probable molars.

Probable molars (Figs. 81N–P; 82F, G) showing wide lingual hypostriae with borders diverging lingually; posterobuccal corners much reduced.

In all upper molariform teeth, cement filling lingual hypostriae and central crescentic valleys; lingual hypostriae on P^4 and M^1 to M^3 crossing nearly half width of tooth and persisting, although becoming shallower, with wear.

MEASUREMENTS

P_3, AP(6), mean 1.41 ± .06, OR 1.19–1.63, CV 10.0; BL tal (6), mean 1.59 ± .08, OR 1.23–1.85, CV 12.6. dP_4, AP(1), 1.94; BL tri (1), 1.58; BL tal (1), 1.67. Lower molariform teeth, *AP(27), mean 1.61 ± .02, OR 1.41–1.83, CV 6.8; BL tri (28), mean 1.78 ± .03, OR 1.53-2.07, CV 7.9; BL tal (29), mean 1.27 ± .02, OR 1.10–1.47, CV 7.9. dP^4, AP(1), 1.23. ?P^4, AP(4), mean 1.25, OR 1.17–1.36; BL(3), mean 2.64, OR 2.29–2.99. ?M^1 or M^2, AP(7), mean 1.24 ± .04, OR 1.10–1.45, CV 8.9; BL(7), mean 2.63 ± .08, OR 2.29–2.86, CV 8.0.

DISCUSSION

Hesperolagomys fluviatilis is clearly a close relative of *H. galbreathi* from the Pliocene of Nevada. *H. fluviatilis* and *Russellagus vonhofi* greatly expand the fossil record of Mio-Pliocene ochotonids showing unilateral hypsodonty.

Both *H. fluviatilis* and *Russellagus vonhofi* are far more common at the Kleinfelder Farm Locality than are related ochotonids at other described localities. *Hesperolagomys* and *Russellagus* predominate over *Hypolagus* at the Kleinfelder Farm Locality, and *Oreolagus*, an ochotonid with unrooted molars more common in other deposits, is absent.

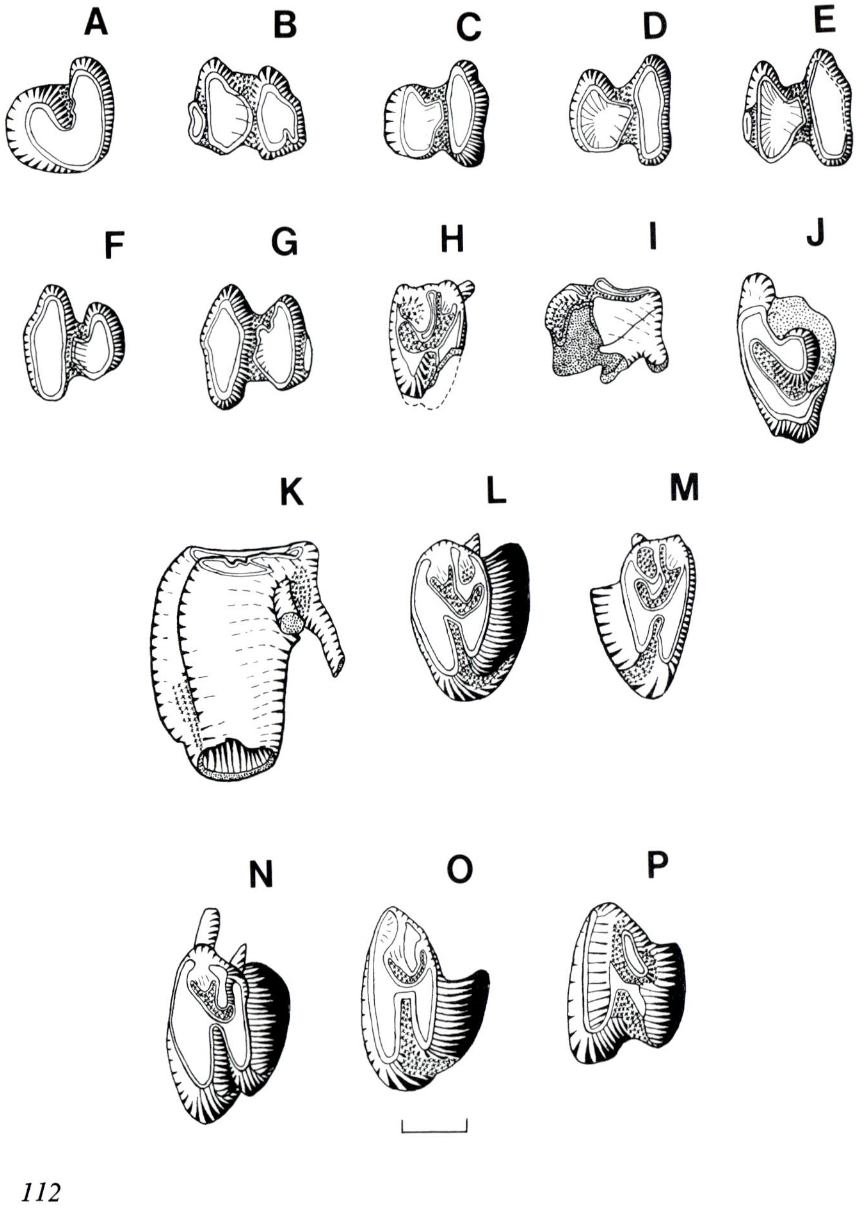

Fig. 81 *Hesperolagomys fluviatilis*, cheek teeth. Scale 1 mm.

A. RP_3, ROM 7306 (paratype), occlusal aspect, anterior to left, buccal to top of page.

B. LdP_4, ROM 7308, occlusal aspect, anterior to right, buccal to top of page.

C. L. lower molariform tooth, ROM 7312, occlusal aspect, anterior to right, buccal to top of page.

D. L. lower molariform tooth, ROM 7314, occlusal aspect, anterior to right, buccal to top of page.

E. L. lower molariform tooth, ROM 7320 (holotype), occlusal aspect, anterior to right, buccal to top of page.

F. R. lower molariform tooth, ROM 7326, occlusal aspect, anterior to left, buccal to top of page.

G. R. lower molariform tooth, ROM 7337, occlusal aspect, anterior to left, buccal to top of page.

H. RdP^4, ROM 7340, occlusal aspect, anterior to right, buccal to top of page.

I. Same specimen as in H, anterior aspect, buccal to right, occlusal to top of page.

J. RP^3, ROM 7339, occlusal aspect, anterior to right, buccal to top of page.

K. ?LM^1 or M^2, ROM 7348 (paratype), posterior aspect, buccal to right, occlusal to top of page.

L. ?LP^4, ROM 7345 (paratype), occlusal aspect, anterior to left, buccal to top of page.

M. ?RP^4, ROM 7352 (paratype), occlusal aspect, anterior to right, buccal to top of page.

N. Same specimen as in K, occlusal aspect, anterior to left, buccal to top of page.

O. ?LM^1 or M^2, ROM 7343 (paratype), occlusal aspect, anterior to left, buccal to top of page.

P. ?LM^1 or M^2, ROM 7346, occlusal aspect, anterior to left, buccal to top of page.

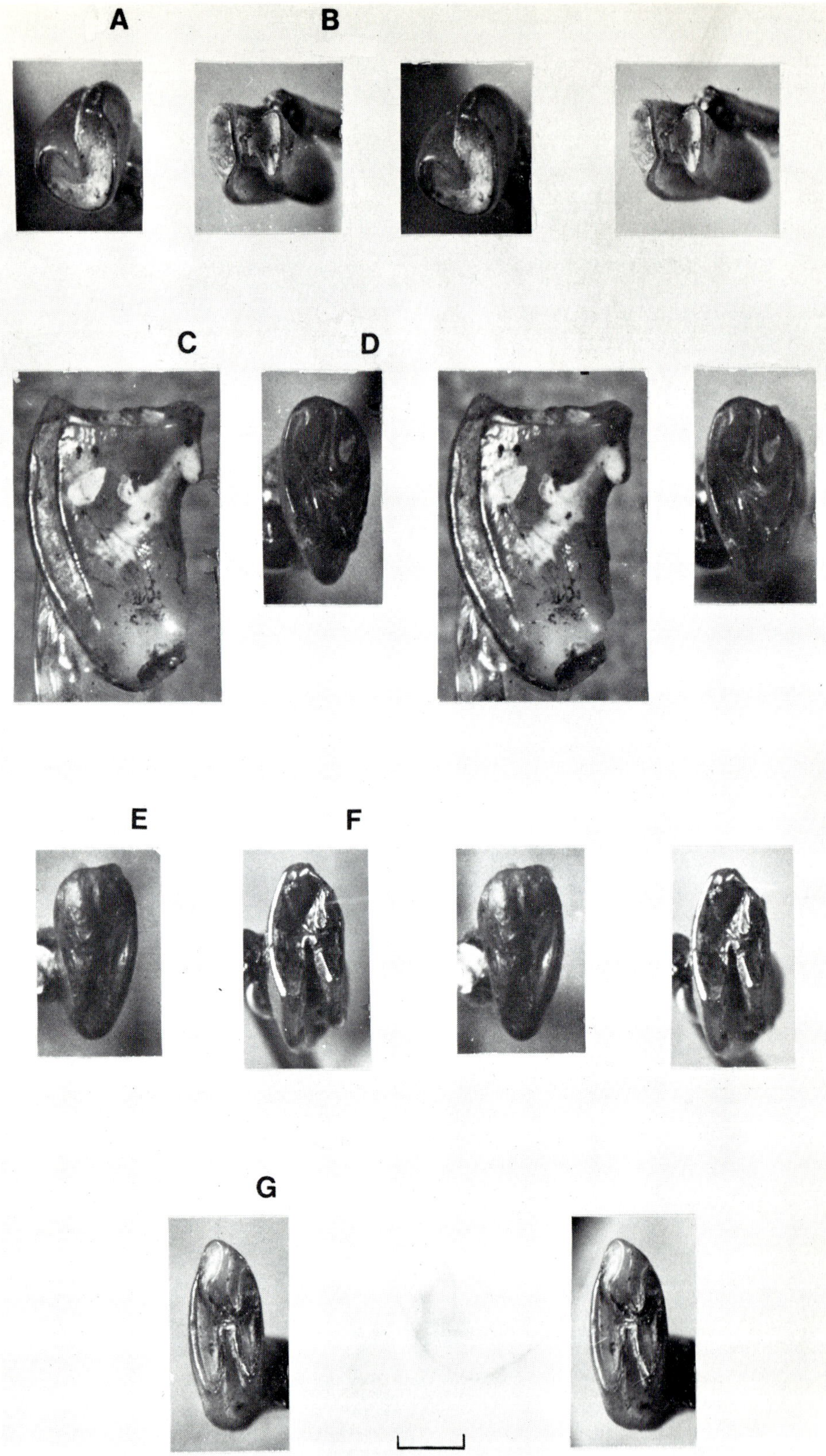
A
B
C
D
E
F
G

Fig. 82 *Hesperolagomys fluviatilis*, cheek teeth. Scale 1 mm.

A. RP_3, ROM 7306 (paratype), occlusal aspect, anterior to left, buccal to top of page.

B. L. lower molariform tooth, ROM 7320 (holotype), occlusal aspect, anterior to right, buccal to top of page.

C. ?LP^4, ROM 7345 (paratype), posterior aspect, buccal to right, occlusal to top of page.

D. Same specimen as in C, occlusal aspect, anterior to left, buccal to top of page.

E. ?RP^4, ROM 7352 (paratype), occlusal aspect, anterior to right, buccal to top of page.

F. ?LM^1 or M^2, ROM 7348 (paratype), occlusal aspect, anterior to left, buccal to top of page.

G. ?LM^1 or M^2, ROM 7343 (paratype), occlusal aspect, anterior to left, buccal to top of page.

Russellagus Storer, 1970a

TYPE SPECIES. *Russellagus vonhofi* Storer, 1970a

HORIZON. Barstovian of Saskatchewan.

DIAGNOSIS OF GENUS. Cheek teeth hypsodont but rooted. Lower jaw with large mental foramen beneath trigonid of M_1, smaller foramen beneath trigonid of P_4. Ventral border of horizontal ramus straight below cheek teeth, curving ventrally to angle. Talonids of lower molariform teeth narrower, longer anteroposteriorly than trigonids, sending forward wide, buccally plicate processes to meet trigonids. Upper cheek teeth unilaterally hypsodont, columns convex lingually with two small buccal roots and one large lingual root. Upper cheek teeth with persistent, cement-filled, crescentic valley and short lingual hypostria traversing one-fourth or less of width of crown and ending lingual to crescentic valley.

Russellagus vonhofi Storer, 1970a

Figs. 83, 84, 85

HOLOTYPE. L. ramal fragment with LP_4, LM_1 (ROM 7384).

PARATYPES. LP^4 (ROM 7399), LM^1 or M^2 (ROM 7409).

REFERRED SPECIMENS. LdP_3 (ROM 7355, 7356), RdP_3 (ROM 7357), LP_3 (ROM 7358–7361), RP_3 (ROM 7362, 7620), LdP_4 (ROM 7363), RdP_4 (ROM 7364, 7365), L. lower molariform tooth (SMNH P558; ROM 7366–7375, 7417), R. lower molariform tooth (ROM 7376–7383, 7418), LP^2 (ROM 7385, 7386), RdP^3 (ROM 7387), LP^3 (ROM 7388), RP^3 (ROM 7389–7391), LdP^4 (ROM 7392, 7393), RdP^4 (ROM 7341, 7394–7398), LP^4 (ROM 7400–7402), RP^4 (ROM 7342, 7403–7405), LM^1 or M^2 (ROM 7406–7408, 7410, 7414, 7415), RM^1 or M^2 (ROM 7411–7413).

Fig. 83 *Russellagus vonhofi*, cheek teeth. Scale 1 mm.

A. LdP_3, ROM 7355, occlusal aspect, anterior to right, buccal to top of page.
B. LP_3, ROM 7358, occlusal aspect, anterior to right, buccal to top of page.
C. RdP_4, ROM 7365, occlusal aspect, anterior to left, buccal to top of page.
D. LP_4, LM_1, ROM 7384 (holotype), occlusal aspect, anterior to right, buccal to top of page.
E. R. lower molariform tooth, ROM 7376, occlusal aspect, anterior to left, buccal to top of page.
F. LP^2, ROM 7385, occlusal aspect, anterior to left, buccal to top of page.
G. RdP^3, ROM 7387, occlusal aspect, anterior to right, buccal to top of page.
H. LdP^4, ROM 7392, occlusal aspect, anterior to left, buccal to top of page.
I. Same specimen as in H, anterior aspect, buccal to left, occlusal to top of page.
J. RdP^4, ROM 7394, occlusal aspect, anterior to right, buccal to top of page.
K. RdP^4, ROM 7395, occlusal aspect, anterior to right, buccal to top of page.
L. LP^3, ROM 7388, occlusal aspect, anterior to left, buccal to top of page.
M. LP^4, ROM 7399 (paratype), anterior aspect, buccal to right, occlusal to top of page.
N. Same specimen as in M, occlusal aspect, anterior to left, buccal to top of page.
O. LP^4, ROM 7400, occlusal aspect, anterior to left, buccal to top of page.
P. LM^1 or M^2, ROM 7409 (paratype), occlusal aspect, anterior to left, buccal to top of page.
Q. LM^1 or M^2, ROM 7415, occlusal aspect, anterior to left, buccal to top of page.

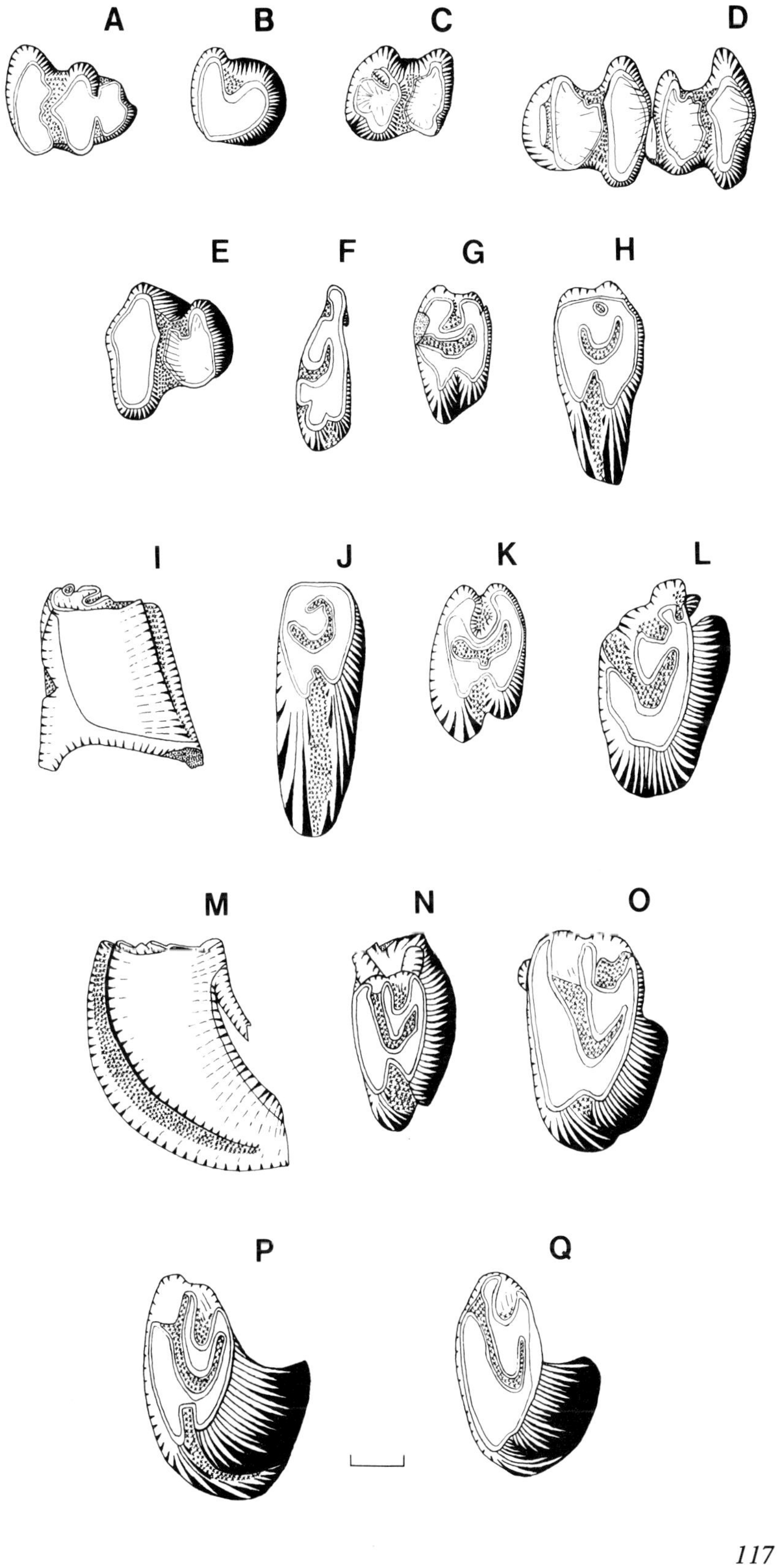
A
B
C
D
E
F
G
H
I
J
K
L
M
N
O
P
Q

TENTATIVELY REFERRED. RP_3 (ROM 7620).

TYPE LOCALITY. Kleinfelder Farm Locality, Lsd. 16, Sect. 31, Twp 2, Rge 30, W. 2nd Merid., near Rockglen, Saskatchewan.

DIAGNOSIS OF SPECIES. As for genus.

DESCRIPTION

Three specimens of dP_3 available, two well worn; ROM 7355 (Fig. 83A) showing union of anterior and central lophids before that of central and posterior lophids, typical of leporids; tooth too small to be referred to *Hypolagus*, must be regarded as unusual individual variant until additional specimens of dP_3 recovered.

P_3 lacking in the holotypic jaw, isolated specimens existing; wider than long, with elongate narrow trigonid (Fig. 83B); cement-filled buccal valley separating talonid and trigonid, extending across half width of tooth; no lingual folds; no cement present on lingual surface of crown; tentatively referred specimen wider than definitely referable specimens.

dP_4 talonid (Figs. 83C, 84A) similar to that of permanent molariform teeth except for absence of buccal plications; large lingual projection, set off by deep anterobuccal and shallow posterolingual valley, occurring in trigonid; no sign of posterolophid as seen in *Hesperolagomys fluviatilis*.

P_4 and M_1, preserved in the holotype (Figs. 83D, 84B), showing many of features of ?*Desmatolagus schizopetrus* (Dawson, 1965, fig. 18); talonids narrower, longer anteroposteriorly than trigonids, sending out broad, lingually rounded, buccally plicate protrusions to trigonids; P_4 and M_1 of holotype showing posterolophids, destroyed by wear sooner in M_1 than in P_4.

Other lower molariform teeth showing different variations in buccal plication of talonid; in ROM 7376 (Fig. 83E) and Dawson's figured lower molariform tooth of ?*D schizopetrus* (1965, fig. 18), trigonid proportionally narrower buccolingually than in P_4–M_1 of ROM 7384 (Figs. 83D, 84B); such specimens possibly third lower molars, which usually have proportionally narrower talonids than other lower molariform teeth.

Diagnostic features of lower jaw shown in holotype (Fig. 84C); mental foramina beneath trigonids of P_4 and M_1; ventral border deepening sharply at angle.

Anterior loph of P^2 (Fig. 83F) narrower transversely, longer anteroposteriorly than wide, short posterior loph; cement-filled crescentic valley almost entirely within anterior loph, opening broadly anterad; lingual hypostria very short, cement-filled; short, cement-filled anterolingual valley in anterior loph.

Third and fourth deciduous upper premolars (Figs. 83G–K; 84D, E) triple-rooted, but unilateral hypsodonty present, base of enamel being horizontal (Figs. 83I, 84E), with buccal roots projecting below enamel margin; occlusal surface narrower than in permanent upper molariform teeth, broadening toward base; amphicone lying anterobuccally on crown, with cement-filled valley (Figs. 83G, K) or basin (Figs. 83H, J; 84D) buccal to it; cement-filled crescentic valley enclosing lingual end of amphicone; cement-filled lingual hypostria traversing one-third width of crown toward lingual end of crescentic valley; third upper deciduous premolar differing from fourth chiefly in smaller size.

In P^3 (Fig. 83L) as in P^2, anterior loph narrower and longer of lophs; cement-filled crescentic valley almost entirely within anterior loph, opening anterad; lingual hypostria consisting of indentation of enamel surface; tooth with posterobuccal cement-filled groove. Permanent upper cheek teeth high-crowned but rooted (Figs. 83M, 85A).

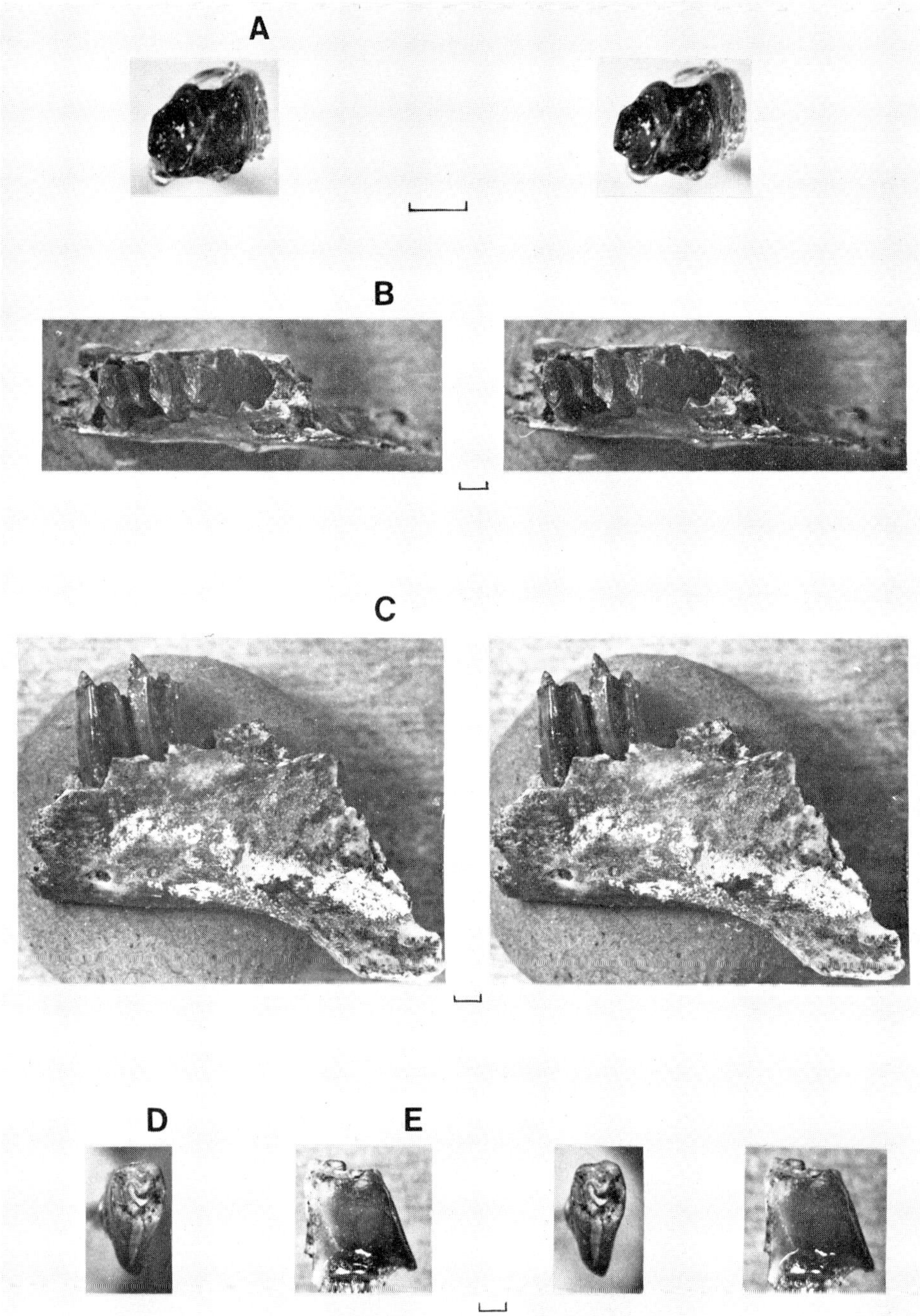

Fig. 84 *Russellagus vonhofi*, cheek teeth. Scales 1 mm.

A. RdP_4, ROM 7365, occlusal aspect, anterior to left, buccal to top of page.

B,C. L. mandible with LP_4, LM_1, ROM 7384 (holotype), anterior to left.

B. Occlusal aspect, buccal to bottom of page.

C. Lateral aspect, occlusal to top of page.

D,E. LdP^4, ROM 7392.

D. Occlusal aspect, anterior to left, buccal to top of page.

E. Anterior aspect, buccal to left, occlusal to top of page.

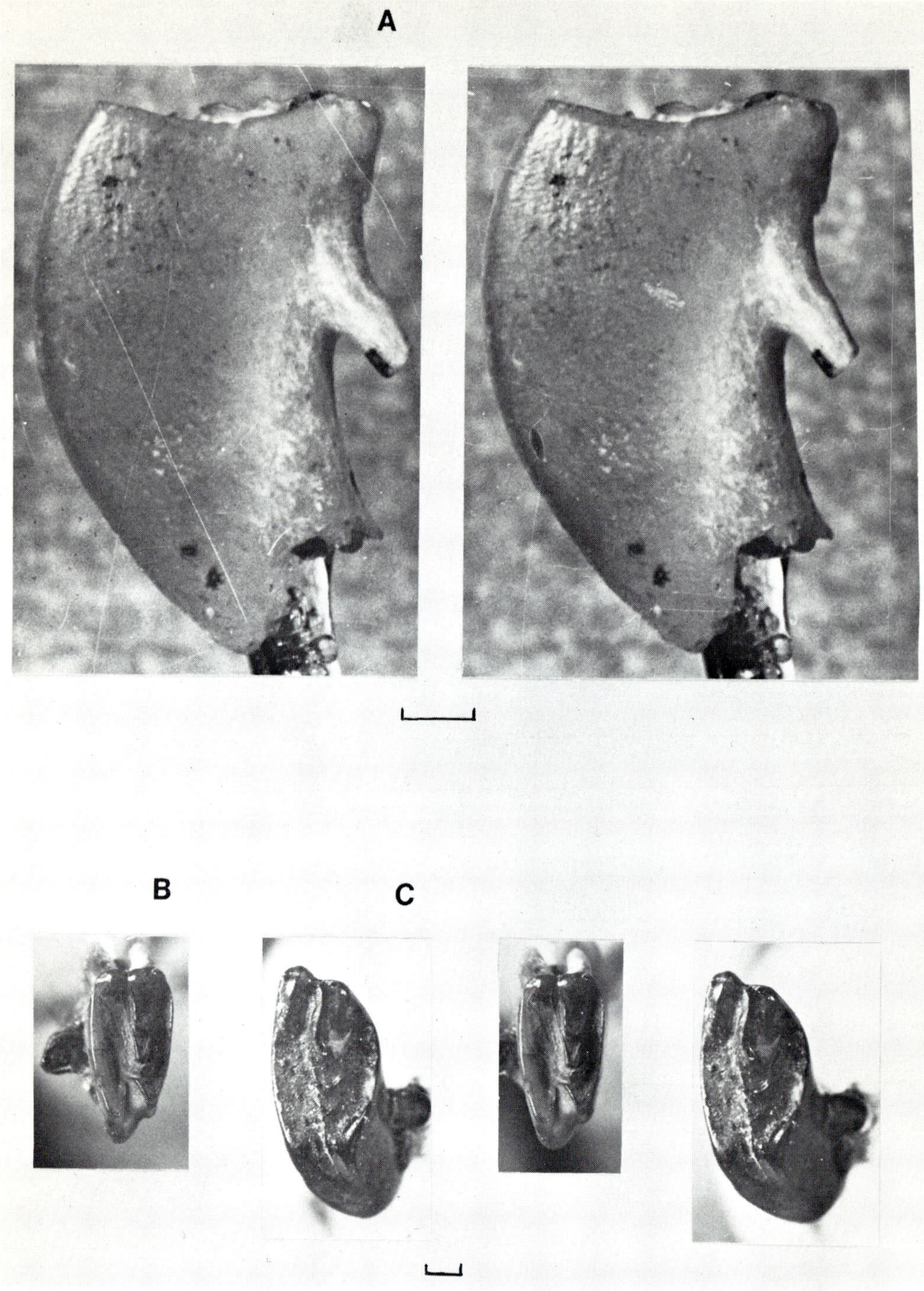

Fig. 85 *Russellagus vonhofi*, cheek teeth. Scales 1 mm.

A. RP^4, ROM 7405, anterior aspect, buccal to right, occlusal to top of page.

B. LP^4, ROM 7399 (paratype), occlusal aspect, anterior to left, buccal to top of page.

C. LM^1 or M^2, ROM 7409 (paratype), occlusal aspect, anterior to left, buccal to top of page.

Anterior loph wider than posterior loph in P^4 (Figs. 83N, O; 85B) and in upper molars (Figs. 83P, Q; 85C), convex anterad in upper molars; as in *Hesperolagomys*, cement-filled crescentic valley of P^4 and M^1 or M^2 forked anteriorly about buccal cuspule (Figs. 83P, 85C, e.g.: further explanation in Fig. 80C); lingual hypostria extending about a quarter of way across surface of crown in little-worn specimens (Figs. 83N, P; 85B, C), ending lingual to lingual end of crescentic valley, but in more worn specimens (Fig. 83O, Q) hypostria extremely short; amphicone, and thus crescentic valley, placed farther lingually than in corresponding teeth of *?Desmatolagus schizopetrus*, *Hesperolagomys galbreathi*, or *H. fluviatilis*; in M^1 or M^2, posterobuccal angle reduced, although not as much as in *H. fluviatilis*, and that of P^4 not reduced.

MEASUREMENTS

dP_3, AP(2), OR 2.16–2.55; BL(2), OR 1.54–1.74. P_3, AP(4), mean 1.56, OR 1.45–1.67; BL(4), mean 1.83, OR 1.67–1.98. Lower molariform teeth, AP(20), mean 2.09 ± .03, OR 1.89–2.31, CV 6.2; BL tri (22), mean 2.30 ± .03, OR 2.02–2.58, CV 5.7; BL tal (22), mean 1.59 ± .03, OR 1.41–1.85, CV 8.2. dP^3, *BL(1), 1.87. P^3, AP(4), mean 1.83, OR 1.67–1.97; BL(4), mean 4.03, OR 3.74–4.40. dP^4, AP(8), mean 1.54 ± .04, OR 1.36–1.67, CV 7.1; *BL(8), mean 2.13 ± .08, OR 1.89–2.68, CV 10.8 P^4, AP(8), mean 1.83 ± .08, OR 1.45–2.07, CV 12.6; BL(8), mean 3.84 ± .18, OR 2.89–4.44, CV 13.0. M^1 or M^2, AP(8), mean 1.65 ± .05, OR 1.41–1.85, CV 9.1; BL(8), mean 4.09 ± .08, OR 3.83–4.49, CV 5.4.

*Measurements of occlusal surface only.

DISCUSSION

Teeth of *Russellagus vonhofi* are smaller than those of *?Desmatolagus schizopetrus*, but are larger than those of *Hesperolagomys fluviatilis* and *H. galbreathi*. Similarities in pattern with teeth of *?D. schizopetrus* are marked, but the two species represent different genera, because the amphicone of *?D. schizopetrus* is closer to the buccal edge of the crown in P^4s and M^1 or M^2s than is the amphicone of *R. vonhofi*, and the lingual hypostria is correspondingly longer in *?D. schizopetrus*. Moreover, neither CM 14785 (P^4) or CM 13567 (M^1 or M^2), specimens of *?D. schizopetrus* with which the type material of *R. vonhofi* was compared, showed any sign of anterior forking of the crescentic valley.

The holotypic jaw of *R. vonhofi* contrasts sharply with the holotype of *Hesperolagomys galbreathi* (Clark, Dawson, and Wood, 1964, fig. 4), in which the mental foramina lie beneath P_3 and M_2, and the ventral border of the horizontal ramus curves slightly dorsally behind the tooth row. Although *Russellagus* is obviously related to *Hesperolagomys*, the two genera are clearly distinct, both in dental pattern and in characters of the lower jaw.

Russellagus vonhofi occurs in greater abundance at the Kleinfelder Farm Locality than does *Hesperolagomys fluviatilis* and is the most common lagomorph there.

Family Leporidae

Hypolagus cf. *H. vetus* (Kellogg), 1910

Figs. 86, 87

REFERRED SPECIMENS. LP_3 (SMNH P552; ROM 7419–7421, 7452), RP_3 (ROM 7422), LP_4 or M_1 or M_2 (SMNH P122—2 specimens, P123, P552, P566; NMC 9429; ROM 7423–7430), RP_4 or M_1 or M_2 (SMNH P552; ROM 7431–7434), L.

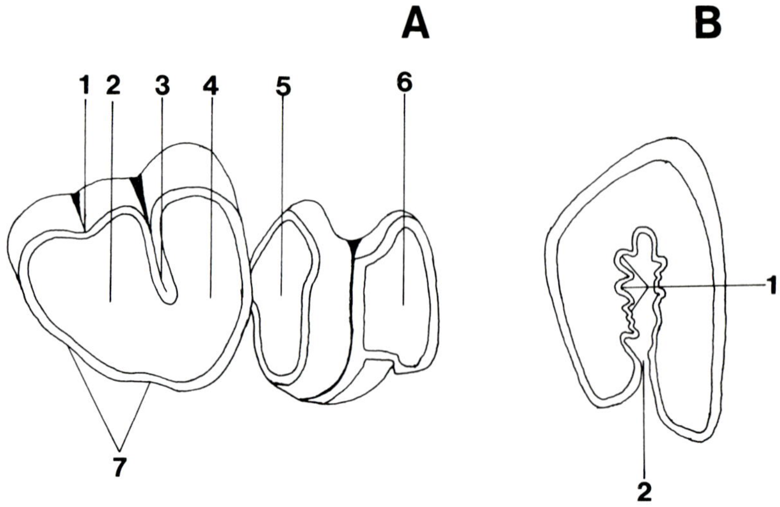

Fig. 86 *Hypolagus* cf. *H. vetus*. Terminology used in description of cheek teeth. Anterior to left, buccal to top of page. Not to scale.

A. RP_{3-4}: 1, anterobuccal re-entrant; 2, trigonid; 3, posterobuccal re-entrant; 4, talonid; 5, trigonid; 6, talonid; 7, lingual indentations.

B. LP^3: 1, crenulations of lingual hypostria; 2, lingual hypostria.

mandibular fragment with LP_{3-4}, LM_{1-2} (NMC 8963), R. mandibular fragment with RP_{3-4}, RM_{1-2} (NMC 8965), LM_3 (SMNH P122; ROM 7435–7438), RM_3 (ROM 7439–7441), LP^2 (ROM 7442, 7443), RP^2 (SMNH P122—2 specimens; ROM 7444–7541), LP^3 (SMNH P122—2 specimens, P552, P558; NMC 9429—2 specimens; ROM 7453–7458), RP^3 (ROM 7459–7463), P^4 or M^1 or M^2 (SMNH P122—5 specimens, P123—3 specimens, P236, P237, P384, P385, P558—2 specimens, P566, SMNH (?); NMC 9392—2 specimens, 9429; ROM 1754, 7464–7471, 7472—13 specimens), LM^3 (ROM 7473, 7474).

DESCRIPTION

P_3 bearing, in many specimens, two very shallow lingual indentations (not visible in Fig. 87A); more posterior indentation facing directly linguad, more anterior anterolinguad; cement never appearing in lingual indentations, or on lingual surface.

Two buccal re-entrants on P_3, both filled with cement; posterobuccal re-entrant running nearly parallel to posterior face of talonid, pointing anterobuccad as in *H. vetus* (Dawson, 1958, fig. 27b); re-entrant showing smooth walls, traversing at least half width of crown; anterobuccal re-entrant, resembling that of *H. vetus*, deeper than in *H. parviplicatus* and *H. apachensis*, sharp, narrow in little-worn specimens (Dawson, 1958, fig. 27b) and more rounded and shallower in worn ones (Dawson, 1958, fig. 27c); cement in Wood Mountain specimens confined to buccal re-entrants, in specimen figured by Dawson (1958, fig. 27b), covering most of buccal surface.

Lower molars of NMC 8963 (Fig. 87A) similar to those figured by Dawson (1958) for *Hypolagus vetus* in shapes of lophids and extent of cement.

Lower jaw (NMC 8963; Fig. 87B) deepening toward posterior end; ascending ramus originating near posterior end of M_2; masseteric ridge strong, especially at ventral end; masseteric ridge crossing buccal surface of ramus, starting at ventral margin and curving onto coronoid process; single mental foramen close to dorsal margin of ramus anterior to third lower premolar.

P^2 (not shown) with same pattern as P^2 of *Hypolagus vetus* (Dawson, 1958, fig. 27a), but cement not covering anterior surface of crown, always confined to lingual hypostria; P^3–M^3 from Wood Mountain Fauna always with cement within lingual hypostria, never extending, as in *H. vetus*, to cover lingual surface.

Diagrams of enamel pattern of upper premolars and molars (Fig. 87C–F) not showing cement, omitting details of column; plications of lingual hypostriae resembling those seen in *H. vetus*, indicating advanced form of *Hypolagus*; P^3 (Fig. 87C) easily distinguishable from other upper molariform teeth on basis of reduced anterobuccal corner; P^4, M^1, and M^2 not distinguishable from each other, cannot be distinguished as left or right; M^3 (not shown) resembling M^3 of *Hypolagus vetus* (Dawson, 1958, fig. 27a) except for extent of cement, similar to that in P^4 to M^2.

MEASUREMENTS

P_3, AP(7), mean 3.1 ± .0, OR 3.0–3.2, CV 3.2; BL tal (7), mean 2.8 ± .1, OR 2.6–3.1, CV 7.1. P_4 or M_1 or M_2, AP(24), mean 2.8 ± .1, OR 2.3–3.3, CV 7.1; BL tri (25), mean 3.1 ± .1 , OR 2.4–3.7, CV 9.7; BL tal (23), mean 2.6 ± .1, OR 2.2–2.9, CV 7.7. M_3, AP(8), mean 1.8 ± .1, OR 1.5–2.0, CV 11.1; BL tri (8), mean 1.7 ± .0, OR 1.5–1.8, CV 5.9. P^2, AP(12), mean 1.3 ± .0, OR 1.2–1.6, CV 7.7; BL(12), mean 2.7 ± .1, OR 2.1–3.1, CV 11.1. P^3, AP(17), mean 2.1 ± .0, OR 1.7–2.4, CV 9.5; BL(17), mean 3.9 ± .1, OR 3.4–4.4, CV 7.7. P^4 or M^1 or M^2, AP(44), mean 2.2 ± .0, OR 1.8–2.6, CV 9.1; BL(42), mean 3.8 ± .1, OR 3.2–4.7, CV 10.5. M^3, AP(2), OR 1.3–1.5; BL(2), OR 2.7–2.8.

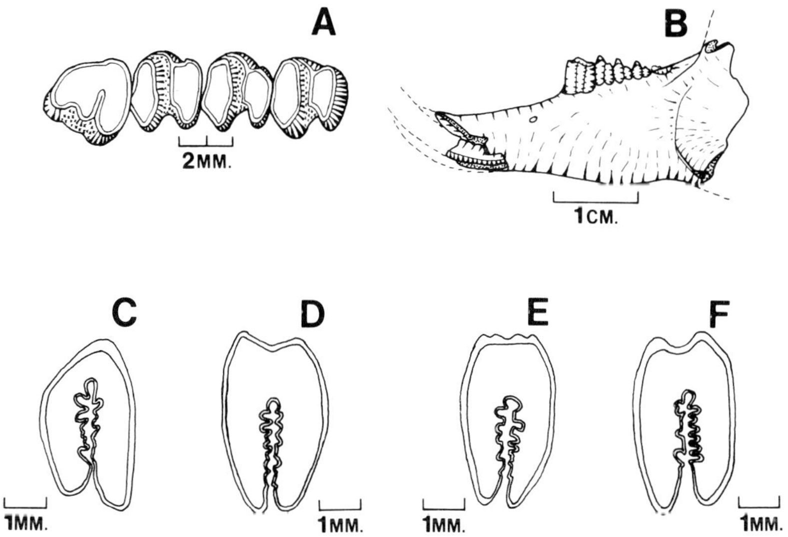

Fig. 87 *Hypolagus* cf. *H. vetus*, cheek teeth and mandible. Scales as indicated.
A,B. L. mandibular fragment with LP_{3-4}, LM_{1-2}, NMC 8963, anterior to left.
A. Teeth in occlusal aspect, buccal to bottom of page.
B. Lateral aspect, occlusal to top of page.
C. LP^3, SMNH P558, occlusal aspect, anterior to left, buccal to top of page.
D-F. P^4 or M^1 or M^2, occlusal aspect, buccal to top of page.
D. SMNH P122.
E. SMNH P123.
F. NMC 9392.

DISCUSSION

The Wood Mountain *Hypolagus* shows only slight differences from the type material of *H. vetus*. Occlusal patterns, as stated above, are very similar, although the extent of cement of P_3 and upper cheek teeth is different. Comparison of Wood Mountain measurements with those of material from the type area of *H. vetus* (Dawson, 1958, tables 12, 13) shows that specimens from the Wood Mountain Formation are somewhat smaller than specimens from the type area of *H. vetus*, although the size ranges overlap.

Many specimens of Barstovian through Hemphillian ages from the Great Plains have been referred to *Hypolagus vetus*. Even a large sample from Wood Mountain does not show differences great enough for the specimens to be referred to a distinct species.

It is probably best to speak of a "*Hypolagus vetus* group", showing slow and limited transition through the late Miocene and the Pliocene, rather than to refer almost every population of *Hypolagus* from the Great Plains of Barstovian-to-Hemphillian age to *H. vetus* proper. Trends within the group include the spread of cement on the cheek teeth, and increase in tooth size, though the differences between the Barstovian and Hemphillian forms are small. It is clear that the *H. vetus* group was a widespread and important taxon in the Barstovian-to-Hemphillian ages of the Great Plains, whatever taxonomic procedures may be applied to its analysis.

Mammalian Faunal List, Wood Mountain Fauna

INSECTIVORA

ERINACEIDAE

Brachyerix cf. *B. incertis* (Matthew), 1924
Parvericius cf. *P. montanus* Koerner, 1940
Amphechinus sp.

TALPIDAE

Domninoides sp.
Scalopoides sp.
Talpidae, gen. et sp. indet.

SORICIDAE

?*Trimylus* sp.
Heterosoricinae, gen. et sp. indet.

CARNIVORA

FELIDAE

Pseudaelurus cf. *P. aeluroides* Macdonald, 1954

MUSTELIDAE

Sthenictis cf. *S. dolichops* Matthew, 1924
Miomustela cf. *M. madisonae* (Douglass), 1929
Plionictis cf. *P. ogygia* (Matthew), 1901
Mustelidae, gen. et sp. indet.

CANIDAE

Tomarctus hippophagus (Matthew and Cook), 1909
Aelurodon cf. *A. saevus* (Leidy), 1858
Leptocyon cf. *L. vafer* (Leidy), 1858

AMPHICYONIDAE

Amphicyon sp.

PROCYONIDAE

Bassariscus cf. *B. parvus* Hall, 1927

URSIDAE

Ursavus sp.

Hemicyon sp.

PROBOSCIDEA

GOMPHOTHERIIDAE

Gomphotheriidae, gen. et sp. indet.

PERISSODACTYLA

EQUIDAE

Merychippus cf. *M. isonesus* (Cope), 1889

Hipparion sp.

Archaeohippus cf. *A. minimus* (Douglass), 1899

Hypohippus cf. *H. osborni* Gidley, 1907

?Anchitherinae, gen. et sp. indet.

Equidae, gen. et sp. indet.

RHINOCEROTIDAE

?*Aphelops* sp.

ARTIODACTYLA

TAYASSUIDAE

Tayassuidae, gen. et sp. indet.

MERYCOIDODONTIDAE

Ticholeptus sp.

CAMELIDAE

Procamelus sp.

PALAEOMERYCIDAE

Blastomeryx cf. *B. elegans* Matthew and Cook, 1909

Cranioceras (*Procranioceras*) cf. *C.* (*P.*) *skinneri* Frick, 1937

cf. *Sinclairomeryx* sp.

ANTILOCAPRIDAE

Submeryceros minor serpentinus Frick, 1937

Merycodus necatus Leidy, 1854

Merycodus sabulonis Matthew and Cook, 1909

Cosoryx(*Paracosoryx*) cf. *C.* (*P.*) *alticornis* Frick, 1937

RODENTIA

MYLAGAULIDAE

Mylagaulus cf. *M. laevis* Matthew, 1902

SCIURIDAE

Spermophilus sp.

Tamias sp.

Sciuropterus sp.

CRICETIDAE

Copemys kelloggae canadensis subsp. nov.

ZAPODIDAE

Megasminthus tiheni Klingener, 1966

Plesiosminthus sp.

EOMYIDAE

Pseudadjidaumo russelli (Storer), 1970a

HETEROMYIDAE

Mookomys sp.

Perognathus saskatchewanensis Storer, 1970a

Perognathoides kleinfelderi Storer, 1970a

Peridiomys borealis Storer, 1970a

Diprionomys cf. *D. agrarius* Wood, 1935

GEOMYIDAE

Lignimus montis Storer, 1970a

CASTORIDAE

Monosaulax cf. *M. pansus* (Cope), 1874

Amblycastor sp.

LAGOMORPHA

OCHOTONIDAE

Hesperolagomys fluviatilis Storer, 1970a

Russellagus vonhofi Storer, 1970a

LEPORIDAE

Hypolagus cf. *H. vetus* (Kellogg), 1910

Age of Fauna

Mammals represented in the Wood Mountain Fauna are sufficiently varied to allow precise determination of the age of the deposits by comparison of the faunal list with those of well-known faunas.

Among the carnivores and ungulates, the following show close similarity to species from the Lower Snake Creek Fauna of Nebraska:

Pseudaelurus cf. *P. aeluroides*
Sthenictis cf. *S. dolichops*
Plionictis cf. *P. ogygia*
Tomarctus hippophagus
Aelurodon cf. *A. saevus*
Leptocyon cf. *L. vafer*
Merychippus cf. *M. isonesus*
Hypohippus cf. *H. osborni*
Blastomeryx cf. *B. elegans*
Submeryceros minor serpentinus
Merycodus necatus
Merycodus sabulonis
Cosoryx (*Paracosoryx*) cf. *C.* (*P.*) *alticornis*

Among the rodents and lagomorphs, *Mylagaulus* cf. *M. laevis*, *Monosaulax* cf. *M. pansus*, and *Hypolagus* cf. *H. vetus* show great similarity to Lower Snake Creek species as well. *Amphicyon* sp. "appears to be closest in size and general dental morphology ... to the Sheep Creek-Lower Snake Creek species placed in *Amphicyon* and cf. *Cynelos*", but shows some peculiarities (R. M. Hunt, Jr., pers. comm., 1972).

Certain large mammals, present in the Wood Mountain Fauna, are missing from the Lower Snake Creek Fauna. *Miomustela* cf. *M. madisonae*, *Bassariscus* cf. *B. parvus*, *Ursavus*, *Hemicyon*, *Hipparion*, *Archaeohippus* cf. *A. minimus*, and *Cranioceras* (*Procranioceras*) cf. *C.* (*P.*) *skinneri*, are typical of deposits somewhat younger than the Lower Snake Creek beds, as are many of the rodents. Thus, although the Wood Mountain Fauna bears similarity to the Lower Snake Creek Fauna, it is significantly younger.

The Niobrara River and Madison Valley Faunas contain the equivalents of *Miomustela*, *Bassariscus*, *Hipparion*, *Archaeohippus*, *Cranioceras*, and some of the rodents of the Wood Mountain Fauna. These faunas represent the uppermost Miocene or lowermost Pliocene.

Closest resemblance of the fine Wood Mountain rodent assemblage is to the

Norden Bridge local fauna (Klingener, 1966, 1968), basal Valentine Formation of Nebraska. *Copemys*, *Megasminthus*, *Plesiosminthus*, *Perognathoides*, and possibly *Tamias*, *Diprionomys*, and *Lignimus* appear to be conspecific in the two faunas. Basal Valentine rocks of Nebraska are the source of the type material of *Cranioceras* (*Procranioceras*) *skinneri*, and the earliest previously known *Hipparion* appears in the basal Valentine (Skinner, pers. comm., 1969). Holman (1970) has also pointed out the great similarity of the herpetofaunas of the Norden Bridge and Wood Mountain Faunas.

The Wood Mountain Fauna, then, can be correlated with well-known deposits from Nebraska. It is just older than the Norden Bridge local fauna and thus is uppermost Barstovian (Upper Miocene), just older than the basal beds of the Valentine Formation.

Acknowledgments

Nearly all the work represented here was incorporated in a dissertation submitted to the University of Toronto (Storer, 1970b). Completion of the project would not have been possible without the constant cooperation of Dr. L. S. Russell, Chief Biologist of the Royal Ontario Museum. He suggested in 1966 that I investigate the Wood Mountain Fauna, and subsequently supported and guided three seasons' field work in the Wood Mountain area, plus research and travel, under National Research Council of Canada Grant A-2685.

Dr. C. S. Churcher, Associate Professor of Zoology at the University of Toronto, served as my committee chairman and gave support and advice in all aspects of my research. The other two committee members, Drs. T. S. Parsons and J. B. Falls, Professors of Zoology, University of Toronto, provided valuable suggestions in the preparation of the thesis.

The directors of the National Museum of Natural Sciences, National Museums of Canada, and the Saskatchewan Museum of Natural History made their Wood Mountain specimens and field data available. I enjoyed the hospitality of the American Museum of Natural History, the Carnegie Museum, the Museum of Paleontology of the University of Michigan, and the Pratt Museum, Amherst College, through the generosity of the respective curators, Drs. M. C. McKenna, C. C. Black, C. W. Hibbard, and A. E. Wood. All allowed me to examine or borrow relevant specimens. I benefited especially from conversations with the above curators, and with Messrs. M. F. Skinner and B. E. Taylor, Dr. R. H. Tedford of the American Museum of Natural History, and Dr. M. R. Dawson of the Carnegie Museum. Dr. R. M. Hunt, of the Department of Anatomy, College of Physicians and Surgeons, Columbia University, compared the skull of *Amphicyon* from the Wood Mountain Formation with material in the collections of the American Museum of Natural History. Thomas H. V. and Pat Vickers Rich of the Department of Geology, Columbia University, made available the draft of their paper on *Brachyerix* (1971), and allowed me to use their information; T. H. V. Rich gave much valuable advice on Wood Mountain erinaceids. Dr. J. H. Hutchison, of the Museum of Paleontology, University of California, Berkeley, gave stimulating comments on the talpids. Dr. J. A. Vonhof generously made available his Masters' (1965) and Ph.D. (1969) theses on Tertiary sands and gravels of Alberta and Saskatchewan; from these much valuable information on the geology of the Wood Mountain Formation was gained.

Particular thanks are due to Mr. John Kleinfelder and family of Rockglen, Saskatchewan, owners of the site of the important Kleinfelder Farm screening locality. Their hospitality has made our quarrying operation enjoyable and productive; no expedition could have more generous and friendly hosts.

Mr. Lawrence Yost of Rockglen has cooperated with Royal Ontario Museum expeditions for some years and has figured in the discovery of many localities. Residents of the entire region from Mankota to Willowbunch, too numerous to mention individually, have furnished information and cooperation and are all in part responsible for the success of Royal Ontario Museum expeditions to the Wood Mountain area.

Throughout the project I have had the cooperation of Dr. A. G. Edmund, Curator of Vertebrate Palaeontology at the Royal Ontario Museum, who gave me the use of laboratory space and equipment. Mr. Gordon Gyrmov, Senior Technician, Department of Vertebrate Palaeontology, gave indispensable technical assistance and advice and did most of the sorting of small bones and teeth from Kleinfelder Farm concentrate. Mr. F. D. Ross, formerly Research Assistant, Department of Vertebrate Palaeontology, assisted in cataloguing. Mr. Anker Odum of the Royal Ontario Museum executed fine illustrations of the skulls of *Amphicyon* and *Merychippus* (Figs. 20, 21, 29, and 30). Mr. Leighton Warren, Chief Photographer, gave great assistance, taking photographs of large specimens and printing others to scale. Messrs. H. C. Smith and R. S. Kidd, of the Provincial Museum of Alberta, read the manuscript and offered valuable suggestions. The manuscript was typed by Miss Elinor Huston, of the Provincial Museum of Alberta.

The referee and editors, through their careful consideration of the manuscript, have helped make this a better publication. Their efforts are greatly appreciated.

Finally, I thank my wife, Barbara, for her patience and understanding throughout the project.

Literature Cited

AMERICAN SOCIETY OF MAMMALOGISTS, COMMITTEE ON NOMENCLATURE

1968 Ground squirrel nomenclature. J. Mamm., vol. 49, no. 3, p. 605.

BLACK, C. C.

1958 A new sicistine rodent from the Miocene of Wyoming. Breviora, no. 86, pp. 1–7.

1961a Fossil mammals from Montana. Pt. 1. Additions to the Late Miocene Flint Creek local fauna. Ann. Carneg. Mus., vol. 36, art. 7, pp. 69–76.

1961b Rodents and lagomorphs from the Miocene Fort Logan and Deep River Formations of Montana. Postilla, no. 48, pp. 1–20.

1963a Miocene rodents from the Thomas Farm local fauna, Florida. Bull. Mus. Comp. Zool. Harv., vol. 128, no. 11, pp . 485–501.

1963b A review of the North American Tertiary Sciuridae. Bull. Mus. Comp. Zool. Harv., vol. 130, no. 3, pp. 109–248.

BLACK, C. C. and A. E. WOOD

1956 Variation and tooth-replacement in a Miocene mylagaulid rodent. J. Paleont., vol. 30, no. 3, pp. 672–684.

BODE, F. D.

1931 Characters useful in determining the position of individual teeth in the permanent cheek-tooth series of merychippine horses. J. Mammal., vol. 12, no. 2, pp. 118–129.

CLARK, J. B., M. R. DAWSON, and A. E. WOOD

1964 Fossil mammals from the Lower Pliocene of Fish Lake Valley, Nevada. Bull. Mus. Comp. Zool. Harv., vol. 131, no. 2, pp. 29–63.

COLLIER, A. J. and W. T. THOM, JR.

1918 The Flaxville gravel and its relation to other terrace gravels of the Northern Great Plains. Prof. Pap. U.S. Geol. Surv., no. 108-J, pp. 179–184.

COPE, E. D.

1874 Notes on the Santa Fé Marls and some of the contained vertebrate fossils. Proc. Acad. Nat. Sci. Philad., vol. 26, pp. 147–152.

1889 A review of the North American species of *Hippotherium*. Proc. Am. Phil. Soc., vol. 26, no. 130, pp. 429–458.

DAWSON, M. R.

1958 Later Tertiary Leporidae of North America. Paleont. Contr. Univ. Kans., art. 6, pp. 1–75.

1965 *Oreolagus* and other Lagomorpha (Mammalia) from the Miocene of Colorado, Wyoming, and Oregon. Univ. Colo. Stud. Ser. Earth Sci., no. 1, pp. 1–36.

DOUGLASS, E.

1899 The Neocene lake beds of western Montana and descriptions of some new vertebrates from the Loup Fork. M.Sc. thesis, University of Montana, 27 pp.

1929 A change of name for *Mustela minor* Douglass. J. Mammal., vol. 10, no. 2, p. 168.

DOWNS, T.

1956 The Mascall fauna from the Miocene of Oregon. Univ. Calif. Publs. Geol. Sci., vol. 31, no. 5, pp. 199–354.

FRASER, F. J., F. H. MCLEARN, L. S. RUSSELL, P. S. WARREN, and R. T. D. WICKENDEN

1935 Geology of southern Saskatchewan. Mem. Geol. Surv. Brch. Can., no. 176, pp. 1–137.

FRICK, C.

1926 The Hemicyoninae and an American Tertiary bear. Bull. Am. Mus. Nat. Hist., vol. 56, art. 1, pp. 1–119.

1937 Horned ruminants of North America. Bull. Am. Mus. Nat. Hist., vol. 69, pp. 1–669.

GIDLEY, J. W.
1907 Revision of the Miocene and Pliocene Equidae of North America. Bull. Am. Mus. Nat. Hist., vol. 23, art. 35, pp. 865–934.

HALL, E. R.
1927 Species of the mammalian subfamily Bassariscinae. Univ. Calif. Publs. Geol. Sci., vol. 16, no. 11, pp. 435–448.
1930 Three new genera of Mustelidae from the later Tertiary of North America. J. Mammal., vol. 11, no. 2, pp. 146–154.

HOFFMEISTER, D. F.
1959 New cricetid rodents from the Niobrara River fauna, Nebraska. J. Paleont., vol. 33, no. 4, pp. 696–699.

HOLMAN, J. A.
1970 Herpetofauna of the Wood Mountain Formation (Upper Miocene) of Saskatchewan. Can. J. Earth Sci., vol. 7, no. 5, pp. 1317–1325.

HOOPER, E. T.
1957 Dental patterns in mice of the genus *Peromyscus*. Misc. Publs. Mus. Zool. Univ. Mich., no. 99, pp. 1–59.

HUTCHISON, J. H.
1966 Notes on some Upper Miocene shrews from Oregon. Bull. Mus. Nat. Hist. Univ. Ore., no. 2, pp. 1–23.
1968 Fossil Talpidae (Insectivora, Mammalia) from the Tertiary of Oregon. Bull. Mus. Nat. Hist. Univ. Ore., no. 11, pp. 1–117.

JAMES, G. T.
1963 Paleontology and nonmarine stratigraphy of the Cuyama Valley Badlands, California. Part 1. Geology, faunal interpretations and systematic descriptions of Chiroptera, Insectivora and Rodentia. Univ. Calif. Publs. Geol. Sci., vol. 45, pp. 1–154.

KELLOGG, L.
1910 Rodent fauna of the Late Tertiary beds at Virgin Valley and Thousand Creek, Nevada. Univ. Calif. Publs. Bull. Dep. Geol., vol. 5, no. 29, pp. 421–437.

KLINGENER, D.
1966 Dipodoid rodents from the Valentine Formation of Nebraska. Occ. Pap. Mus. Zool. Univ. Mich., no. 644, pp. 1–9.
1968 Rodents of the Mio-Pliocene Norden Bridge local fauna, Nebraska. Am. Midl. Nat., vol. 80, no. 1, pp. 65–74.

KOERNER, H. E.
1940 The geology and vertebrate paleontology of the Fort Logan and Deep River Formations of Montana. Am. J. Sci., ser. 5, vol. 38, pp. 837–862.

LEIDY, J.
1854 [On *Bison latifrons*, *Arctodus pristinus*, *Hippodon speciosus*, and *Merycodus necatus*]. Proc. Acad. Nat. Sci. Philad., vol. 7, pp. 89–90.
1858 Notice of remains of extinct Vertebrata from the valley of the Niobrara River, collected during the expedition of 1857, in Nebraska, under the command of Lieut. G. K. Warren, U.S. Top. Eng., by Dr. F. V. Hayden. Proc. Acad. Nat. Sci. Philad., vol. 10, pp. 20–29.

LINDSAY, E. H.
1972 Small mammal fossils from the Barstow Formation, California. Univ. Calif. Publs. Geol. Sci., vol. 93, pp. 1–104.

LOVE, J. D.
1961 Split Rock Formation (Miocene) and Moonstone Formation (Pliocene) in central Wyoming. Bull. U.S. Geol. Surv., no. 1121-I, pp. 1–39.

LUNDELIUS, E. L., JR.
1960 *Mylohyus nasutus*, long-nosed peccary of the Texas Pleistocene. Tex. Meml. Mus. Bull., no. 1, pp. 1–40.

MACDONALD, J. R.
1948 The Pliocene carnivores of the Black Hawk Ranch fauna. Univ. Calif. Publs. Geol. Sci., vol. 28, no. 3, pp. 53–80.

1954 A new *Pseudaelurus* from the Lower Snake Creek fauna of Nebraska. J. Paleont. vol. 28, no. 1, pp. 67–69.

MCGREW, P. O.

1938 The Burge fauna; a Lower Pliocene mammalian assemblage from Nebraska. Univ. Calif. Publs. Geol. Sci., vol. 24, no. 11, pp. 309–328.

MATTHEW, W. D.

1901 Fossil mammals of the Tertiary of north-eastern Colorado. Mem. Am. Mus. Nat. Hist., n.s., vol. 1, no. 7, pp. 355–447.

1902 A horned rodent from the Colorado Miocene, with a revision of the *Mylagauli*, beavers, and hares of the American Tertiary. Bull. Am. Mus. Nat. Hist., vol. 16, art. 22, pp. 291–310.

1924 Third contribution to the Snake Creek fauna. Bull. Am. Mus. Nat. Hist., vol. 50, art. 2, pp. 59–210.

MATTHEW, W. D. and H. J. COOK

1909 A Pliocene fauna from western Nebraska. Bull. Am. Mus. Nat. Hist., vol. 26, art. 7, pp. 361–414.

MERRIAM, J. C.

1916 Tertiary vertebrate fauna from the Cedar Creek Mountain region of western Nevada. Univ. Calif. Publs. Bull. Dep. Geol., vol. 9, no. 13, pp. 161–198.

OSBORN, H. F.

1904 New Miocene rhinoceroses with revision of known species. Bull. Am. Mus. Nat. Hist., vol. 20, art. 27, pp. 307–326.

1918 Equidae of the Oligocene, Miocene and Pliocene of North America, iconographic type revision. Mem. Am. Mus. Nat. Hist., n.s., vol. 2, pt. 1, pp. 1–330.

PATTON, T. H.

1969 Miocene and Pliocene artiodactyls, Texas Gulf Coastal Plain. Bull. Fla. St. Mus. Biol. Sci., vol. 14, no. 2, pp. 115–226.

REED, K. M.

1962 Two new species of fossil talpid insectivores. Breviora, no. 168, pp. 1–6.

RENSBERGER, J. M.

1971 Entoptychine pocket gophers (Mammalia, Geomyoidea) of the Early Miocene John Day Formation, Oregon. Univ. Calif. Publs. Geol. Sci., vol. 90, pp. 1–209.

REPENNING, C. A.

1967 Subfamilies and genera of the Soricidae. Prof. Pap. U.S. Geol. Surv., no. 565, pp. 1–74.

RICH, T. H. V. and P. V. RICH

1971 *Brachyerix*, a Miocene hedgehog from western North America, with a description of the tympanic regions of *Paraechinus* and *Podogymnura*. Am. Mus. Novit., no. 2477, pp. 1–58.

ROMER, A. S.

1966 Vertebrate Paleontology. 3d ed. Chicago, University of Chicago Press. 468 pp.

ROSE, B.

1916 Wood Mountain-Willowbunch coal area, Saskatchewan. Mem. Geol. Surv. Brch. Can., no. 89, pp. 1–103.

RUSSELL, L. S.

1933 A new species of *Merychippus* from the Miocene of Saskatchewan. Can. Fld. Nat., vol. 47, no. 1, p. 11.

1937 New and interesting mammalian fossils from western Canada. Trans. R. Soc. Can., ser. 3, vol. 30, sec. 4, pp. 75–80.

1950 The Tertiary gravels of Saskatchewan. Trans. R. Soc. Can., ser. 3, vol. 44, sec. 4, pp. 51–59.

1965 Tertiary mammals of Saskatchewan. Part I: The Eocene fauna. Life Sci. Contr., R. Ont. Mus., no. 67, pp. 1–33.

1972 Tertiary mammals of Saskatchewan. Part II: The Oligocene fauna, non-ungulate orders. Life Sci. Contr., R. Ont. Mus., no. 84, pp. 1–97.

SCHULTZ, C. B. and C. H. FALKENBACH

1941 Ticholeptinae, a new subfamily of oreodonts. Bull. Am. Mus. Nat. Hist., vol. 79, art. 1, pp. 1–105.

SHOTWELL, J. A.

1967a *Peromyscus* of the Late Tertiary in Oregon. Bull. Mus. Nat. Hist. Univ. Ore., no. 5, pp. 1–35.

1967b Late Tertiary geomyoid rodents of Oregon. Bull. Mus. Nat. Hist. Univ. Ore., no. 9, pp. 1–51.

1968 Miocene mammals of southwest Oregon. Bull. Mus. Nat. Hist. Univ. Ore., no. 14, pp. 1–67.

SKINNER, M. F. and B. E. TAYLOR

1967 A revision of the geology and paleontology of the Bijou Hills, South Dakota. Am. Mus. Novit., no. 2300, pp. 1–53.

STERNBERG, C. M.

1930 Miocene gravels in southern Saskatchewan. Trans. R. Soc. Can., ser. 3, vol. 24, sec. 4, pp. 29–30.

STIRTON, R. A.

1935 A review of the Tertiary beavers. Univ. Calif. Publs. Geol. Sci., vol. 23, no. 13, pp. 391–458.

STORER, J. E.

1969 An Upper Pliocene neohipparion from the Flaxville gravels, northern Montana. Can. J. Earth Sci., vol. 6, no. 4, pt. 1, pp. 791–794.

1970a New rodents and lagomorphs from the Upper Miocene Wood Mountain Formation of southern Saskatchewan. Can. J. Earth Sci., vol. 7, no. 4, pp. 1125–1129.

1970b The Wood Mountain fauna: an Upper Miocene mammalian assemblage from southern Saskatchewan. Ph.D. thesis, University of Toronto. 400 pp.

1973 The entoptychine geomyid *Lignimus* (Mammalia:Rodentia) from Kansas and Nebraska. Can. J. Earth Sci., vol. 10, no. 1, pp. 72–83.

VANDERHOOF, V. L. and J. T. GREGORY

1940 A review of the genus *Aelurodon*. Univ. Calif. Publs. Geol. Sci., vol. 25, no. 3, pp. 143–164.

VONHOF, J. A.

1965 Tertiary gravels and sands in southern Saskatchewan. M.Sc. thesis, University of Saskatchewan. 99 pp.

1969 Tertiary gravels and sands in the Canadian Great Plains. Ph.D. thesis, University of Saskatchewan. 287 pp.

VOORHIES, M. R.

1969 Taphonomy and population dynamics of an early Pliocene vertebrate fauna, Knox County, Nebraska. Contr. Geol. Univ. Wyo., Spec. Pap. no. 1, pp. 1–69.

WEBB, S. D.

1969 The Burge and Minnechaduza Clarendonian mammalian faunas of north-central Nebraska. Univ. Calif. Publs. Geol. Sci., vol. 78, pp. 1–191.

WILLIAMS, M. Y. and W. S. DYER

1930 Geology of southern Alberta and southwestern Saskatchewan. Mem. Geol. Surv. Brch. Can., no. 163, pp. 1–160.

WILSON, R. W.

1960 Early Miocene rodents and insectivores from northeastern Colorado. Paleont. Contr. Univ. Kans., art. 7, pp. 1–92.

WOLFE, J. L. and J. N. LAYNE

1968 Variation in dental structures of the Florida mouse, *Peromyscus floridanus*. Am. Mus. Novit., no. 2351, pp. 1–7.

WOOD, A.E.

1935 Evolution and relationships of the heteromyid rodents with new forms from

the Tertiary of North America. Ann. Carneg. Mus., vol. 24, art. 7, pp. 73–262.
1936a Fossil heteromyid rodents in the collections of the University of California. Am. J. Sci., ser. 5, vol. 32, pp. 112–119.
1936b Geomyoid rodents from the Middle Tertiary. Am. Mus. Novit., no. 866, pp. 1–31.
1937 Additional material from the Tertiary of the Cuyama Basin of California. Am. J. Sci., ser. 5, vol. 33, no. 193, pp. 29–43.